EMPOWERING THE BEGINNING TEACHER OF MATHEMATICS IN

# MIDDLE SCHOOL

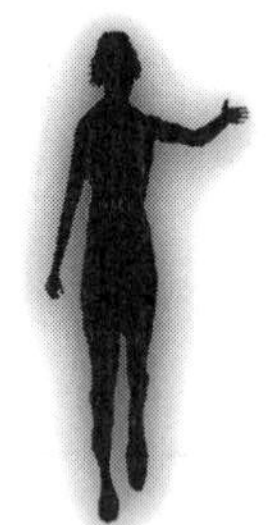

EMPOWERING THE BEGINNING TEACHER OF MATHEMATICS

*A series edited by Michaele F. Chappell*

EMPOWERING THE BEGINNING TEACHER OF MATHEMATICS IN

# MIDDLE SCHOOL

Edited by

Michaele F. Chappell
*Middle Tennessee State University*
*Murfreesboro, Tennessee*

Tina Pateracki
*Jasper County Schools, Ridgeland South Carolina*

NCTM | NATIONAL COUNCIL OF TEACHERS OF MATHEMATICS

1906 Association Drive, Reston, VA 20191-1502
(703) 620-9840; (800) 235-7566; www.nctm.org



Library of Congress Cataloging-in-Publication Data

Chappell, Michaele F.
Empowering the beginning teacher of mathematics in middle school / edited by Michaele F. Chappell, Tina Pateracki.
p. cm. — (Empowering the beginning teacher of mathematics)
Includes bibliographical references.
ISBN 0-87353-560-X
1. Mathematics—Study and teaching (Middle school) I. Pateracki, Tina II. Title. III. Series.
QA11.2.C43 2004
510'.71'2—dc22

2004004366

The National Council of Teachers of Mathematics is a public voice of mathematics education, providing vision, leadership, and professional development to support teachers in ensuring mathematics learning of the highest quality for all students.

The publications of the National Council of Teachers of Mathematics present a variety of viewpoints. The views expressed or implied in this publication, unless otherwise noted, should not be interpreted as official positions of the Council.

Printed in the United States of America

# CONTENTS

# INDUCTION AND MENTORING OF NEW TEACHERS

## Position

The National Council of Teachers of Mathematics believes that school systems and universities must assume the shared responsibility for the sustained professional support of beginning teachers by providing them with a structured induction and mentoring program. This effort must include opportunities for further development of mathematics content, pedagogy, and management strategies. Association with a trained mentor who has a strong background in mathematics, mathematics pedagogy, and classroom practice is crucial to this program.

## Background and Rationale

"Before It's Too Late," a report of the National Commission on Science and Mathematics Teaching in the 21st Century, recommended that teachers be initiated into the profession through induction programs. In many school settings, new mathematics teachers who may not have strong mathematics content knowledge are isolated and given little support and content-specific professional development. In these circumstances, their students are not afforded the learning opportunities and quality instruction advocated by the Council.

The retention of new teachers continues to be a problem, contributing to the overall shortage of mathematics teachers. A research review by Yvonne Gold found that 30 to 37 percent of new teachers leave the profession within their first five years ("Beginning Teacher Support: Attrition, Mentoring, and Induction," in *Handbook of Research on Teacher Education*, 2nd ed., edited by John Sikula, Thomas Buttery, and Edith Guyton [New York: Macmillan, 1996], pp. 548–94).

## Recommendations

- School systems should develop structured induction programs that include mentoring.
- University teacher-preparation programs should serve as a partner with school districts in induction programs by participating in the training of mentors, continuing communication with their graduates, and serving as a resource.
- Mentor teachers should be provided with significant and consistent training and be given additional remuneration or release time for their services.
- Schools should set aside time specifically for the collaborative efforts of the beginning teacher and the mentor.
- District and school administrators should recognize the added demands on beginning teachers and their mentors and should be sensitive in making teaching assignments.
- Districts and universities should offer professional development that includes a strong focus on content knowledge, pedagogical knowledge, pedagogical content knowledge, and a knowledge of *Principles and Standards for School Mathematics* and its applications to the classroom.

(August 2002)

# PREFACE

## Yes, you have made a great career choice as a teacher!

—*Shirley M. Frye*

Teaching is a rewarding profession. As you embark on what may be the most important adventure of your life—that is, the process of teaching students mathematics—take comfort in the words of Shirley Frye, NCTM Past President, spoken at the Beginning Teachers Conference held at the NCTM Eastern Regional Conference in Boston, Massachusetts, in November 2002. Have confidence in the knowledge that you have acquired from your educational experiences thus far. Exercise patience with yourself as you strive to achieve higher levels of competence and reach proficiency.

As you begin your journey as a teacher of mathematics, you are likely to encounter challenges—both inside and outside the classroom—that will seem to overshadow the perceived rewards of teaching. Realize that in your early years of teaching mathematics, you will probably have a "large learning agenda" (Feiman-Nemser 2003, p. 27) that may require you to gain more knowledge about the content you are teaching and how best to present it to your students. This agenda may also require that you learn more about the norms of teaching among your colleagues and in your school community. Although obstacles will surface during your early years of teaching, you should view them as unique learning opportunities that enable you to refine your existing skills and polish your daily practices as you progress along the path of mastery in your new career.

To assist you in this process, the Editorial/Author Panel for the Needs of Mathematics Teachers Beginning Their Careers has compiled this middle school volume to help you reach your full potential as an effective teacher of mathematics, thereby improving the mathematics learning of the students who will be the recipients of your instruction throughout your career. The Empowering the Beginning Teacher of Mathematics series contains three books geared specifically toward elementary, middle, and high school teachers of mathematics. These books have been written both *for* you and *to* you. Several authors present their discussions objectively, with the beginning teacher in mind, but many share their wisdom and insights as if they were conversing with you over a cup of tea. We hope that this level of familiarity will set the tone for your use of this volume.

Our initial charge and primary goal was to develop a resource to which beginning teachers of mathematics could refer and one that they would use often while attending to the many demands of the classroom and the teaching profession in general. We all know that each academic year brings new faces and new demands to the classroom, at times making even veteran teachers feel like beginners again. Thus, we anticipate that this volume may also serve as a source of inspiration for both beginning teachers and their more experienced colleagues.

The Panel has aimed to produce a unique resource that highlights varied contributions in six broad categories: (1) professional growth, (2) curriculum and instruction, (3) classroom-level assessment, (4) classroom management and organization, (5) equity, and (6) school and community. To us, these categories represent the essential domains to which beginning teachers of mathematics must give immediate attention during the early years to establish a firm foundation in the classroom and to pave the way for a long tenure in mathematics teaching. In each category, individual contributions take on different formats, including featured articles, related thematic ideas, bulleted lists of tips and advice, personal testimonies, quick notes that shed light on specific topics, and quotable thoughts that can be stated best only by teachers. Journal-like pages are included at the end of each section for you to make notes and add your personal ideas, stories, tips, or advice to which you can refer in subsequent years or share with colleagues.

From the onset of the writing project, we were careful to avoid producing a volume that mirrors the numerous resources already available in teacher journals and related books. We certainly encourage you, as a beginning teacher of mathematics, to make full use of these resources as you seek to learn more about the situations you encounter in your first few months in the classroom. Our desire, however, is that you do more than merely "read and shelve" this publication. We hope that you keep it close at hand during your early years as a teacher and that you think of it as an *active* resource—one that becomes an integral part of your teaching regimen—in your search for solutions to issues and problems, not solely mathematical, that are sure to arise in your classroom or school during your beginning years.

Numerous people have made possible the production of the books in the Empowering the Beginning Teacher of Mathematics series. I especially offer my gratitude to the other members of the Editorial/Author Panel for their innovative, diligent, and focused work:

- Jeffrey M. Choppin, University of Rochester, Rochester, New York
- Tina Pateracki, Jasper County Schools, Ridgeland, South Carolina
- Jenny Salls, Washoe County School District, Sparks, Nevada
- Jane F. Schielack, Texas A&M University, College Station, Texas
- Sharon Zagorski, Milwaukee Public Schools, Milwaukee, Wisconsin

Throughout this project, the members of the Panel have contributed countless hours reviewing, editing, and crafting supporting segments to prepare this entire volume for *you*—the beginning teacher of mathematics in the middle school. I also wish to acknowledge Harry Tunis, our staff liaison at the National Council of Teachers of Mathematics (NCTM), for his unwavering support and guidance, as well as the production staff of NCTM for assistance in the editing and production of this work. Finally, I wish to thank the authors, who have contributed to this effort as a response to their own desire to see you develop into an enthusiastic, effective classroom practitioner.

Our hope in producing this volume is that you "emerge from [your] first few years of teaching [mathematics] feeling empowered, supported, and capable in all roles of the classroom teacher" (Renard 2003, p. 64). You can help yourself in this endeavor by recognizing the multifaceted roles and responsibilities that teachers of mathematics assume during their beginning years. Moreover, as NCTM's position statement about new teachers suggests, you should, if possible, take part in a high-quality induction or mentoring program. Ultimately, you should position yourself to reach out to your future colleagues who will enter the field after you and share the ideas that you learn from this volume, other resources, and your own experience.

Yes, you have made a great career choice as a teacher of mathematics in the middle school. Now we urge you to enjoy your journey!

*Michaele F. Chappell*
*Series Editor*

# INTRODUCTION

*Tina Pateracki*
*Michaele F. Chappell*

As a teacher of middle school mathematics, you will extend the mathematical ideas and understandings that your students bring from their elementary school experience. In so doing, you will prepare them for a more in-depth study of mathematics in high school and postsecondary education. One imperative toward meeting that goal is that you create an environment in which these students come to view mathematics as useful and exciting. Doing so calls for understanding the unique characteristics of the middle school child, as well as having a deep understanding of the mathematics content taught in the middle grades and appropriate pedagogy for teaching it.

Middle school students are entering adolescence, which for them is a time of tremendous physical, intellectual, and emotional change. They can be quite introspective as they develop their own self-image. They are intensely curious and exhibit a strong willingness to learn things that they consider useful. Middle schoolers prefer active over passive learning. Relevant, interesting, and challenging mathematics lessons will not only address their needs but will also help them develop an appreciation for mathematics. Middle-grades students are sensitive to criticism and value the opinion of their peers; they tend to have a keen sense of fairness. At the same time, they are making crucial decisions about themselves as learners that can influence their attitudes, motivation, and participation in mathematics for the remainder of their lives. As a teacher of middle school mathematics, you will want to capitalize on the characteristics of students at this level. Equally important, you should establish norms and procedures that support the learning of mathematics by *all* students.

As you begin your career as a mathematics teacher in the middle school, you undoubtedly have questions and trepidations about what to expect. Thus, for you, we have compiled a series of articles that offer guidance and advice as you plan for and experience your first year. In the first section, "Professional Growth," are articles that encourage reflection on your teaching and assist with your own professional growth. The second section, "Curriculum and Instruction," presents suggestions and tools to help you plan innovative lessons that incorporate meaningful discourse and active learning. No lesson would be complete without an assessment of what students know and are able to do as a result of their learning. Hence the third section, "Classroom Assessment," presents ways to assess students beyond the traditional classroom means. The areas that can most influence whether your students will maximize their mathematics learning in your classroom are classroom management and organization. Accordingly, in section 4, strategies for these two crucial topics are discussed. Perhaps at no other time in a student's school career is diversity seen as such an issue. Students come into the middle grades not only with diverse ethnic, religious, and cultural backgrounds but also with varying levels of maturity, achievement, and mathematics backgrounds. Creating equitable classroom so that *all* students can learn mathematics is essential at this level; to this end, section 5, "Equity," discusses several aspects of this important issue. Finally, developing partnerships with the families of your students and the community at large can yield many rewards. You will find suggestions on how to begin this process in section 6, "School and Community."

Your first year likely will present moments of both triumph and challenge. Celebrate first your triumphs, as these moments of joy will sustain you through the latter and will create memories that endure. Yet this book is written to help you address the obstacles and answer the questions that are sure to arise in your years as a beginning teacher. As you read through it, jot down questions and notes in the margins; talk with your mentor and colleagues about them; and reflect on your successes and less-than-successes. Allow this book to become your journal as you chronicle your journey of becoming an effective teacher of mathematics in the middle school.

# Professional Growth

Your first years of teaching will be challenging and rewarding—and stressful. You, as a new entrant to the profession, are expected to assume the same responsibilities as a twenty-year veteran, including everything from operating the copy machine to teaching reform-based curriculum. You may need to adapt and develop lessons, discover how to use new materials, determine the most effective classroom-management skills, and meet the needs of diverse students. As a teacher, you should continuously experiment with new methods and try to learn from your successes and mistakes. We should all strive to do so! As you face the realities of teaching, you may wonder where to turn to continue your professional growth. The following paragraphs suggest several novel resources for beginning the professional development that should take place throughout your career.

## Self-Assessment

Reflecting on your own teaching is a vital step in your growth. Analyze your lessons, and think about what went well and what you might change. This reflection can lead to improved lesson planning and teaching practices. Keeping a journal is one way to record your reflections. Some questions to ponder include the following:

- What did I do in my lesson?
- What were the goals for my lesson?
- Did I accomplish my goals?
- What anticipated challenges did my students face during the lesson?
- What were some different ways of thinking that I observed as my students worked on a given task?
- How might I revise this lesson in the future?

## Your Colleagues

Collaborating with fellow teachers is another way to grow professionally. Find colleagues who are knowledgeable and willing to share ideas that work. Of course, not all strategies that are effective for an experienced teacher will work for you. Be selective. Seek out new ideas and resources. Ask questions. And remember to share with others what works for you!

## Support Groups

Many schools and districts offer formal induction programs and support groups for beginning teachers. Often groups of new teachers meet weekly or monthly to share common concerns and successes. Mentoring programs are also becoming more popular. Consider selecting and working with a mentor. Be willing to seek out more formal support groups. You do not have to face all your challenges alone!

## Professional Journals and Organizations

Keep up with current practices and issues in education by reading professional journals. Find time each month to read one or two articles that interest you. Local, state, and national organizations hold annual meetings, academies, and workshops to help you grow professionally. Learn more about what conferences are offered in your area, and attend a conference or workshop to see how valuable such gatherings can be.

## Coursework

You may have graduated only recently, but more coursework may be in your future. In the months to come, you might consider expanding your knowledge of mathematics content and pedagogy, as well as classroom practice, through some form of teacher education. Take time to investigate programs, and talk to others in your field about appropriate coursework for the topics in which you are interested.

Although this entire book is intended to provide support and ideas for your growth as a teacher of mathematics, this first section deals specifically with your professional growth. As you read the pages ahead, consider the following questions:

- How do I continue to develop my skills as a teacher to improve the abilities of my students?
- Where do I turn for feedback and advice on my teaching?
- How do I make my first year of teaching successful for me and for my students?

We believe that the habits of reflection you develop as you read this book and think about these questions will serve you well as you seek to achieve growth in your teaching career.

Photograph by Makoto Yoshida. 

# Four Crucial Insights for First-Year Teachers of Mathematics

*Steve Leinwand*

In light of the typical absence of collaborative mechanisms to "learn the ropes" of teaching—particularly the teaching of mathematics—I offer this set of four crucial insights for thriving as a beginning teacher. These insights are gleaned from extensive discussions with first- and second-year teachers and from the review of many professional performance portfolios. Moreover, they arise out of a desire to extend the "wisdom of practice" garnered by experienced teachers but shared too infrequently with those who are new to the profession.

## Insight 1 Just because it worked once …

One of the most discouraging realities of teaching is that even though a certain approach worked first period, you have no guarantee that it will work fifth period. Conversely, just because something was a disaster this year or during your last-period class, you have no reason not to try it again next year or during a different period. The fact is that classroom dynamics and the distinctive personality of each student and class are often far more powerful determinants of the success or failure of a lesson than your plans. The excitement of teaching—even after years of practice—arises from the unique set of circumstances that you face every year with every new class and from the ongoing struggle to refine and modify your methods. This perspective helps you overcome the daily frustrations and the inevitable classes that "bomb."

## Insight 2 Mistakes happen

Another aspect of teaching that is seldom discussed is how rare the "perfect class" is. In fact, you learn quickly that teaching a forty-five-minute class without making at least two mistakes is almost impossible. One mistake is usually a careless mathematical error made because you are thinking several steps ahead. Sometimes your students will catch the mistake, and sometimes the error sits on the chalkboard unnoticed until the classroom erupts in a disagreement about the final answer. The second mistake is usually pedagogical and results from calling on the wrong student at the wrong time or assigning the wrong problem at the wrong time. Either error is sure to engender confusion. When a supervisor or principal is in the room or when you are using technology, the likelihood of mistakes increases significantly. Once you realize that such mistakes are typical in all classes and almost inevitable during any given class, you can begin to shift your perspective and see most mistakes not as embarrassments but as welcome learning opportunities for both you and your students.

## Insight 3 Do not try to work alone

The professional isolation of teachers is among the most serious impediments to improving practice and developing teaching skills. Unfortunately, most teachers practice their craft behind closed doors, minimally aware of what their colleagues are doing and usually unobserved and undersupported. Perpetuating this debilitating culture is irrational. Instead, you should realize that remedies for nearly every teaching obstacle reside among your colleagues if you are only willing to ask. As lonely as you may feel in your career, you are not alone. Through direct communication with colleagues, online interaction with other teachers, or the range of professional development opportunities offered locally and regionally, support and helpful suggestions are readily available.

## Insight 4 If you do not occasionally feel inadequate, you are probably not doing the job

Just think about what you are being asked to do: teach in distinctly different ways from how you were taught mathematics; use hardware and software that did not exist a few years ago; make much more frequent use of group work; focus as much on problems, communication, and applications as on skills and procedures; teach groups of students that are far more heterogeneous than those of your predecessors; and assess understanding in more authentic ways. Feeling overwhelmed by this torrent of change is neither a weakness nor a lack of professionalism. It is an entirely rational response. A reasonable perspective is that an occasional sense of inadequacy is both inevitable and typical and should be channeled into stimulating the ongoing growth and learning that characterize the true professional.

Keep these insights in mind as you begin your career as a teacher; doing so may save you the sometimes-painful experience of learning them on your own. •

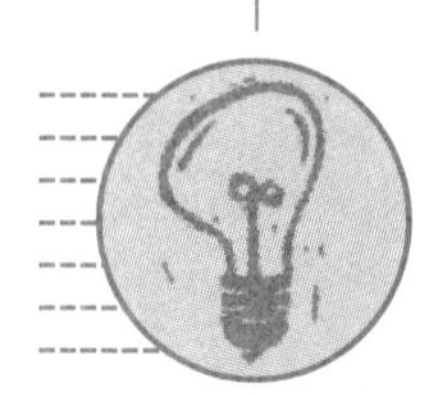

## Keeping a Proper Perspective about Your Students

*Darren A. Cooper*

One of the greatest professional lessons I ever learned was from a student named Ollie Gary*. This student did everything imaginable to be disruptive and to get under my skin, and he succeeded. I love children—and mathematics—but at the time, I hated Ollie Gary. Other children's misbehavior seemed to be youthful mischief or bravado, but to me, Ollie's antics were mean-spirited and intolerable. After one particularly trying day, I was fuming over something Ollie had done. Then I had a revelation: I taught twenty-nine students who adored me and frequently made their feelings evident; I was crazy to let the one remaining child irritate me. For the rest of the year, Ollie Gary never upset me again. Oh, he tried, but I was able to dismiss his annoying behavior. My resolution to tie my emotions to the many appreciative students I teach remains valuable to me; quite often, I have "that one student" in a class who is difficult in some way, yet I cope.

I have no greater joy in my life than teaching mathematics. My wish for beginning mathematics teachers is that you find the same delight and passion that I find in my teaching. I have three pieces of advice for you: Love the mathematics, love the children, and never let one student get you down.

* A pseudonym

I WISH I HAD KNOWN

## Top Ten Things I Wish I Had Known When I Started Teaching

*Cynthia Thomas*

10. Not every student will be interested every minute. No matter how much experience you have or how great you are at teaching, you will encounter times in the classroom when no student is interested! The solution is to change your tone of voice, move around the room, or switch from lecturing to some other activity. Maybe you can even use a manipulative to increase the students' understanding and, possibly, their level of interest.

9. If a lesson is going badly, stop. Even if you have planned a lesson and have a clear goal in mind, if your approach is not working—for whatever reason—stop! Regroup and start over with a different approach, or abandon your planned lesson entirely and go on to something else. At the end of the day, be honest with yourself as you examine what went wrong and make plans for the next day.

8. Teaching will get easier. Maybe not tomorrow or even next week, but at some point in the year, your job *will* get easier! Try to remember your first day in the classroom. Were you nervous? Of course; all of us were. See how much better you are as a teacher already? By next year, you will be able to look back on today and be amazed at how much you have learned and how much easier so many aspects of teaching are!

7. You do not have to volunteer for everything. Do not feel that you always have to say yes each time you are asked to participate. Know your limits. Practice saying, "Thank you for thinking of me, but I do not have the time to do a good job with another task right now." Of course, you must accept your responsibility as a professional and do your fair share, but remember to be realistic about your limits.

6. Not every student or parent will love you. And you will not love every one of them, either! Those feelings are perfectly acceptable. We teachers are not hired to love students and their parents; our job is to teach students and, at times, their parents as well. Students do not need a friend who is your age; they need a facilitator, a guide, a role model for learning.

5. You cannot be creative in every lesson. In your career, you *will* be creative, but for those subjects that do not inspire you, you can turn to other resources for help. Textbooks, teaching guides, and professional organizations, such as NCTM, are designed to support you in generating well-developed lessons for use in the classroom. When you do feel creative and come up with an effective and enjoyable lesson, be sure to share your ideas with other teachers, both veterans and newcomers to the profession.

4. *No one* can manage portfolios, projects, journals, creative writing, and student self-assessment all at the same time and stay sane! The task of assessing all these assignments is totally unreasonable to expect of yourself as a beginning teacher. If you want to incorporate these types of exercises into your teaching, pick one for this year and make it a priority in your classroom. Then, next year or even the year after that, when you are comfortable with the one extra assignment you picked, you can incorporate another innovation into your teaching.

3. Some days you will cry, but the good news is, some days you will laugh! Learn to laugh *with* your students and *at* yourself!

2. You will make mistakes. You cannot undo your mistakes, but berating yourself for them is counterproductive. If the mistake requires an apology, make it and move on. No one is keeping score.

1. This *is* the best job on earth! Stand up straight! Hold your head high! Look people in the eye and proudly announce, "I am a teacher!"

**Above all else, remember: "Those who can, teach. Those who cannot, do something *far less* important!"**

## Choosing and Working with a Mentor

*Sharon Zagorski*

New teachers continuously search for support, resources, and ideas during their first years in the profession to make sense of the realities of teaching. One important source of support, the use of mentor teachers, is becoming more prevalent both nationally and internationally. Finding and working with a mentor is a good idea for almost any new teacher. The following lists highlight important questions to consider when you are looking for someone to fulfill this role in your professional life.

### What qualities should I look for in a mentor?

- A knowledgeable teacher who is committed to the profession
- A teacher who has a positive attitude toward the school, colleagues, and students and is willing to share his or her own struggles and frustrations, avoiding the naysayer who constantly complains in staff meetings
- A teacher who is accepting of beginning teachers, showing empathy and acceptance without judgment
- A teacher who continuously searches for better answers and more effective solutions to problems rather than believes that he or she already has the only right answer to every question and the best solution to every problem
- A teacher who leads and attends workshops and who reads or writes for professional journals
- An open, caring, and friendly individual who has good communication skills
- Someone who shares your teaching style, philosophy, grade level, or subject area
- A teacher who is following the path you want to follow, someone with whom you can relate and with whom you share mutual respect
- Someone who is aware of his or her own biases and opinions and encourages you to listen to advice but also to form your own opinions

## What should I expect from an effective mentor?

- A mentor allows you to talk without interruptions and listens for your sake.
- A mentor maintains confidentiality in your discussions and interactions.
- A mentor helps you explore options, set goals, and attempt to do things your way, using your strengths and personality.
- A mentor builds on your strengths and avoids trying to transform you into a teacher clone using his or her style.

## What are my responsibilities as a new teacher working with a mentor?

- Welcome the mentor's interest and concern.
- Realize that both partners can gain from the relationship.
- Realize that mutual respect, trust, and openness are the foundations for achieving success.
- Avoid a passive role; take the initiative in your own development by specifying your needs, soliciting feedback, and using the feedback without viewing it as criticism or an evaluation.
- Have realistic goals and expectations for what can be accomplished. Be open and sincere.
- Communicate any difficulties and concerns as clearly as possible. Be willing to discuss failures, as well as successes. Understand that learning comes from an examination of both.
- Follow through on commitments, and seek help when necessary. Asking for help is a sign not of weakness but of strength.
- Be honest with your mentor about important feelings. Contribute ideas and a variety of options for overcoming difficulties.

## What are the benefits of having a mentor?

- Having a mentor gives you an opportunity to learn from an experienced teacher, who shares his or her personal knowledge, experiences, and insights.
- A mentor who helps you understand and cope with written and unwritten rules will ensure that you are quickly assimilated into the school culture.
- Working with a mentor gives you the chance to test ideas, strategies, and tactics in a friendly forum before you try them in a classroom.
- Having a mentor gives you access to coaching and counseling.
- A mentor can help you clarify your career goals by making you aware of local, state, and national professional organizations, thus opening the doors to continuing growth and development.

Your work with a mentor will be as rewarding and successful as you make it. This relationship should serve as a strong foundation for support and future professional growth. •

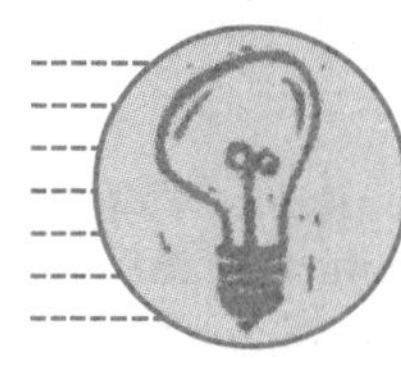

## Josephine's Story: A Constructive Collaboration

*Jane Lincoln Miller*
*Josephine To Tam*

New teachers can use a lot of help! However, giving and accepting such help is often a sensitive issue in the culture of schools (Little 1990) because veteran teachers may be reluctant to offer unsolicited assistance and, similarly, new teachers may be reluctant to reveal their need for help. The potential advantages of asking for and accepting help are many—as Josephine's story reveals. During Josephine's first year of teaching mathematics in a middle school, she sought help from two fellow seventh-grade mathematics teachers, and a constructive collaboration evolved. Here is Josephine's story.

> As I settled in during that first year, I observed my colleagues and asked a lot of questions. The fact that I was willing to approach other teachers and ask for their advice, expectations, and materials helped a lot! Two other seventh-grade mathematics teachers, Dawn L. and Kristine N., and I developed a strong team support group. Initially, I did not have much experience to draw on to anticipate students' needs and potential

pitfalls as they learned mathematics. However, Dawn and Kristine each had at least four years of teaching and were familiar with our seventh-grade mathematics curriculum. We met weekly to plan and to share responsibilities, such as selecting worksheet tasks, copying, and managing computer-related tasks. As the new teacher on the team, I went to all the school system training [sessions] on mathematics and mathematics teaching. The ideas and information that I brought back were incorporated into the lessons we planned together.

In our planning and teaching, we sought a balance between instructing from a constructivist approach with investigations, hands-on activities, and discovery of concepts and a skills-based approach with practice of routine procedures. We asked students a lot of "why" questions. When worksheets and skills were the focus, we tried to link the concepts with practical applications to challenge students to think.

In my second year, Dawn, Kristine, and I began taking advantage of the new staff development substitute-teacher program to make use of uninterrupted time to plan. Planning for several hours at a time allowed us to prepare long-range [lessons] and have a global perspective about curriculum and assessment matters. The three of us were able to talk about strategies, decide how we wanted to teach a particular topic, and come up with practical projects, such as a menu project and a recipe project on proportional reasoning. In planning together we anticipated a lot of student thinking and tackled head-on any mathematics misconceptions students might have [had]. As we continue to collaborate, we develop the flow of our lessons and our whole unit according to our collective experiences.

This constructive collaboration began when Josephine, during her first year of teaching, decided to ask for help. At the time of this writing, this collaboration is going strong in its fourth year. The teachers take full advantage of resources provided by their school system; they have gained invaluable opportunities for mutual support and professional growth. Opportunities for their students to enjoy mathematics and learn it more effectively have been greatly enhanced. Their collaborative model has enormous potential for helping first-year mathematics teachers—and veterans, too!

## Keeping a Professional Journal

*Susan Kyle Arn*

As you begin teaching, keep a good professional journal. Once a month, update this journal by noting any professional development meetings you have attended, presentations you have made, professional organizations you have joined, and so on. With the journal, keep any certificates of attendance or completion you have earned and a copy of your transcripts, along with a copy of your evaluation and your current teaching certificate. You may also want to include articles from the newspaper or professional journals that you value or notes on topics of interest in your career. After a few years of teaching, you will need to create a new journal, but remember to keep the old one.

You will be surprised how much you will add to your journal each month. It will also come in handy when you need to update your resume or you begin to apply for awards and grants. This journal will become one of the most important professional references that you have.

### Remember—Write It Down!

A great way to keep track of plans that did not go well, lessons that took too much time, or specific ideas or exercises that drove a lesson home is by writing them directly on your lesson script. Recalling a good teaching suggestion from a previous year may be difficult unless you take the time to jot it down where you are sure to see it when it is needed.

—*Barbara A. Burns*

# I NEVER LEARNED

## SEVEN THINGS I NEVER LEARNED IN METHODS CLASS

*Margaret R. Meyer*

1. Do not think that students never notice what clothes you wear or when you last cut your hair. They are quite observant about such things because these concerns are very important in their own lives. When building a professional wardrobe, do make the choice of comfort over fashion, especially when you are buying shoes.

2. Do not bore your friends with school stories unless they are teachers, too. A story that is funny to a teacher is often not funny to those in other occupations. Do try to balance your life with friends who work outside of education.

3. Do not take your health for granted when working with children. Keep a box of tissues on your desk, and insist that students use them. Ask students to bring in replacement boxes from home; they are usually happy to do so. Wash your hands frequently.

4. Do not think you will always be twenty-something. Pay attention to saving for your retirement. Take advantage of tax-sheltered savings plans.

5. Do not take too long to recover from your undergraduate degree. Start a graduate program as soon as possible. Doing so will pay off well in the long run.

6. Do not isolate yourself behind your closed door. Find colleagues with whom you can talk, plan, share successes and failures, and continue to grow professionally.

7. Do not ever tell your students how old you are, especially when they ask you directly. Instead, add at least thirty years to your age when answering because that age is how old they really think you are. Do think about retiring when your answer starts to sound believable.

Notes:

2 SECTION

# CURRICULUM AND INSTRUCTION

Teaching involves many decisions, most of which must be made before a lesson begins. These decisions give teachers an opportunity to reflect on the kinds of learning experiences they intend for their students. As a teacher, you may find yourself deciding such issues as what topic should be the focus of a lesson, how to engage students in that topic, what questions to ask students, how to guide discussions to both encourage participation and advance particular mathematical ideas, and what tools or resources to use. These decisions are important because they influence the kinds of learning opportunities and views of mathematics that your students will have. In this section, you will find ideas to guide you as you contemplate these questions.

This section has three major themes: (1) Planning, (2) Questioning and Discourse, and (3) Instructional Tools and Resources. The themes reflect multiple and mostly distinct facets of the complex decision-making you will face as a teacher. As you design and carry out your instructional plans, you will need to consider and reflect on all three areas to teach mathematics for understanding.

## Planning

Most likely, planning will occupy a major portion of your time during your beginning years of teaching. To carry out the day-to-day functions of teaching and to be effective in your eyes and in the eyes of your colleagues and administrators, you must understand the scope and nature of the curriculum that you are teaching and the best methods for implementing that curriculum. Planning requires that you consider both the short-term view—what you plan to accomplish in a given lesson—and the long-term view—what should be the lasting learning outcomes of your teaching, such as instilling a mathematical disposition and an ability to solve problems in your students. In this section, you will find guidance to assist you in planning, including how to determine what you are responsible for teaching, what classroom policies are appropriate and effective, and what instructional strategies to use.

## Questions and Discourse

As a teacher, you serve as the representative of the larger academic and mathematical communities, and you control the flow of information to your students and the depth to which they think about mathematical ideas. One important component that defines your role is the kinds of questions you ask students during instruction. These questions can serve to assess students' knowledge and to initiate students into mathematical discussions. Because different forms of questions serve distinct purposes, you will need to consider and balance your short-term and long-term goals in determining whether to ask a question that requires a brief factual response or one that requires an extended response in which students must explain their thinking. Reading this section will help you determine the types of questions to ask, both in planning your lesson and in conducting it with your students.

## Instructional Tools and Resources

While planning your lessons, you will need to identify what tools and resources, such as technology and manipulatives, to incorporate and how best to use them. For example, you may need a system for distributing and accounting for manipulatives and calculators or advice on how to design lessons that include the use of computer software. Lessons that integrate calculators, computers, and manipulatives may require a good deal of time; however, such lessons tend to offer different avenues through which students can engage in mathematics. As a beginning teacher, you may want to implement tools and resources gradually as you learn how their use can enhance your students' mathematical understanding.

# ABCs for Teachers: Answers to Beginning Concerns

*Gladis Kersaint*
*Edward Mooney*

As a new teacher, you have many new challenges ahead of you—students to get to know; school and district policies to learn and implement; and undoubtedly, questions to answer about teaching, learning, and assessing mathematics. To help you meet these challenges, here are a few responses to "frequently asked questions" posed by beginning and change-of-career teachers.

## What am I responsible for teaching?

Not all beginning or change-of-career teachers are given copies of the mathematics curriculum guide or framework that they are responsible for teaching. Often, teachers are simply handed the textbooks that are being used for the courses they will teach. If you do not have a copy of the curriculum guide or framework for your courses, request one from the principal or appropriate school staff member. Typically, your state's department of education will outline the mathematics curriculum; however, your school district may have developed its own curriculum that fits the state's guidelines. Many states now have their curriculum frameworks available online.

## How much of the curriculum will I be expected to teach? How will I cover everything?

Most likely, teaching the required curriculum is feasible, but you will need to put the task into perspective. First, review the curriculum and the textbook. Look at the extensive list of topics that you are to teach. Ask yourself, "Can any of these topics be combined through the use of appropriate instructional strategies or learning experiences?" You may be able to identify a number of creative ways to integrate various topics. With few exceptions, most conventional textbooks should be considered resources for instruction, not the sole basis for instruction. Having reviewed your curriculum, you will be better prepared to decide when using other resources may be more appropriate.

## What should I emphasize when I teach?

What you emphasize will depend on your teaching philosophy and that of your school or district. If you have not done so already, begin to develop a philosophy about what you expect your students to gain from your mathematics instruction. Proceed from those ideas by focusing on the topics that you believe are important. When you know what you want to accomplish in your classroom, determining whether you are reaching your goals will be easier. Compare the importance of completing pages of drill exercises with helping students develop or expand their reasoning skills. With emerging technologies, the focus on procedures alone has become less important. Students need to understand mathematical concepts, know when and how to apply them, and know how to extend and justify their mathematical ideas.

## How do I get my students to cooperate?

Students today are quite savvy and seldom accept actions without appropriate rationales. Talk to your students about what you are trying to achieve in the classroom, and why. Students will know what to expect when they are informed about and understand your goals for them. They must find that the reasons you provide match the concerns that are important to them. This awareness will help students interpret the interactions they experience in your classroom.

Depending on their experiences in other mathematics classrooms, students may initially resist any approach that differs from the norm. The good news is that with time, fortitude, and conviction on your part, students can learn to appreciate their newly developed skills or the strategies that you are encouraging.

## How much homework should I assign?

Assign as much homework as necessary to augment your instruction. That is, decide what you expect from the homework. What information will it provide you? Can you get the same information by assigning only five problems instead of fifty?

As a new teacher, you may easily find yourself falling into "survival mode," in which you become so fixated on the day-to-day occurrences in the classroom that you neglect long-term planning. If you begin with the overall goal in mind, then you can prepare more effectively for the opportunities and challenges that lie ahead.

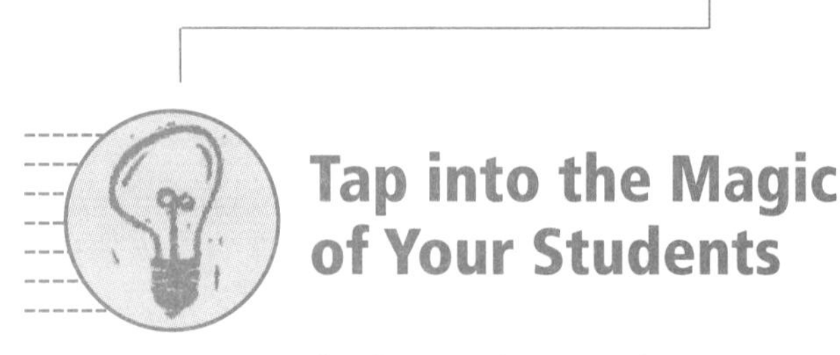

## Tap into the Magic of Your Students

*Claudia Bertolone-Smith*

Here are some ideas on how to find magic in your mathematics teaching:

- Know how each lesson fits into the bigger picture of what you are teaching. For example, if you are teaching a unit about multiplication, each lesson should somehow tie into that goal. Remind and show students how each lesson is related to multiplication.
- Be prepared for the students who will understand a topic right away and those who will need extra help. "I'm done" and "I don't get it" are both valid states of mind when learning mathematics. Make sure that you acknowledge and honor each intellectual state in your classroom, and know how you will help students move on meaningfully.
- If you want magic in your lessons, here is where to find it: in your students. Inside each of them is a curious, investigative, patterning, sorting, and classifying mind. Include them. Ask them what they think. Ask them what they see. Ask them to share their ideas, opinions, and reactions.

## Questions to Ask While Planning an Engaging Mathematics Lesson

*Margaret E. McIntosh*
*Roni Jo Draper*

Lessons that are dull to students are difficult to pull off; and lessons that are dull to teachers are impossible to pull off. To keep your instruction varied and fresh, ask yourself the following questions as you prepare your lessons:

### How can this lesson be a game?

Students enjoy games, and sometimes just the act of playing a game energizes the class. Popular games, such as Pictionary, relay races, bingo, Jeopardy, Wheel of Fortune or hangman, and Concentration, can often be modified for use in the classroom.

### How many different ways can this information be presented?

Students benefit from repetition, especially if the repetition comes in a variety of forms. Look for multiple ways to present and reinforce the topic at hand.

### What novel activity could accompany this lesson?

Novel activities, such as projects, give students opportunities to apply their knowledge. Projects can also motivate students to want to learn.

### How can students get a multisensory experience with this lesson?

Mathematics instruction should include all the senses—touch, sight, hearing, smell, and taste. For example, when students use orange peels to demonstrate that the surface area of a sphere is four times the area of one great circle of the sphere, they will not forget the concept or the experience.

### What can students write about?

Many forms of writing can be used in the classroom to accomplish different purposes. For example, students can take notes using words, numbers, diagrams, symbols, or pictures—whatever means works for them. Learning logs give students the opportunity to do a small amount of writing in a short period of time. Students can also engage in longer writing assignments, such as composing essays, creative stories or problems, poetry, or children's books.

### What kind of connections can be made?

Always seek ways to make connections among various topics in mathematics. For example, algebra can be applied to probability just as easily as it can be applied to problems in percents. You may also look for ways to make connections between topics in mathematics and other disciplines. For example, using the metric system, computing percent error, and graphing and interpreting data are frequently required in science and social studies.

### What reading assignment could enhance this lesson?

Research indicates that when students are given two passages to read on the same topic, they show better comprehension of the second passage and remember more about it. Even children's books may be useful when you are looking for reading material to reinforce students' learning. •

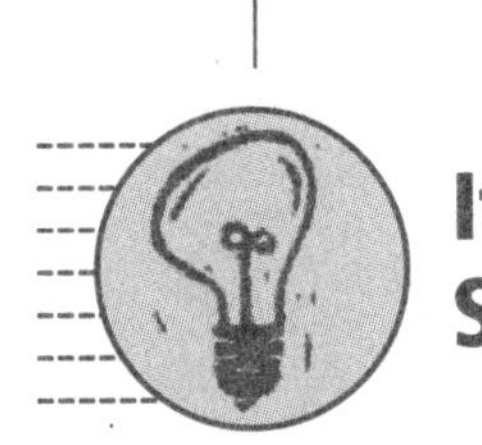

## It Takes Only a Spark

*Angiline Powell*
*Christine Zhou*

It was my first year of teaching mathematics to eighth-grade students. I had recently graduated from college with a degree in mathematics education, and I was excited about teaching. I truly believed in my heart that with enough effort and time, I could wipe out mathematical ignorance in my lifetime. However, I was not prepared for the overwhelming challenge of facing 130 students who came in droves, asking questions and making demands that had nothing to do with mathematics. My students wanted me to sign field-trip forms, absence forms, and early dismissal forms. They forgot their pencils, homework, paper, and books. On top of that, the beginning of class was often plagued with outbursts. What was I going to do?

A friend suggested a five-minute warm-up to start my classes. After giving the suggestion careful thought the night before, I developed what I called the "Planet Nog" activity by generating a set of theme-based critical-thinking questions for the purpose of the warm-up.

At the end of the grading period, I reflected on the benefits of the warm-ups. I had started using them to help cope with the large number of students and their many demands. Then I noticed that some students who did not normally participate in mathematics class seemed to enjoy responding to the questions. They begged for them. If I did not prepare the warm-up, my students were so disappointed that they offered to find problems in my critical-thinking books. As time passed, I realized other benefits of the warm-ups. I noticed that the number of student outburst decreased. More students began to think logically and critically. Additionally, I was able to incorporate more problem solving, communication, and reasoning during a class period. So what started as a spark for me evolved into a fire for my students.

## What? Using Literature to Engage Mathematics Students with the Standards?

*Cindie Heinrich Donahue*
*Denisse R. Thompson*

Picture this scene in a middle school mathematics classroom. The teacher is standing in front of the classroom. All the students are listening. You can tell that they are listening because their heads are up, there are facing the teacher, they are quiet, and they are sitting still. Sound impossible? Want to know what the teacher is doing? The teacher is reading a children's literature book to the students. No, this scenario is not fictitious; it really happens!

### Fitting Stories into the Curriculum

How can children's stories fit into a mathematics curriculum? With all the skills that need to be taught and all the highs-takes testing taking place, how does a teacher even find time to read children's books to select appropriate ones to use to teach a particular topic? Testing has put a great deal of pressure on both students and teachers to cover a state- or district-mandated curriculum. Many states now have standards-driven testing programs; although many such tests use multiple-choice and short-answer items, they are increasingly incorporating essay or performance tasks in mathematics. Literature can be used to develop tasks that help prepare students for open-ended-assessment questions.

### Using *The Gift of the Magi*

As an illustration, we share highlights of some activities that we did with students before the December holidays, using *The Gift of the Magi* as the opening. The story enabled us to talk about "real giving" and set the tone for the activities that followed.

Each group was given the names of four children along with a list of several presents that each child would like to receive as holiday gifts; the context for the list was the "Angel Tree" program, which provides presents for foster children or children whose parents are incarcerated. The students were asked to look through store advertisements and catalogues to price the gifts listed for each child. The students then had to find the total cost of the gifts for each child as well as the average cost per gift. They had to group the gifts into four different categories and make bar graphs showing the total cost and the average cost for each. Finally, they were asked to respond to a newspaper reporter who wants to learn more about the types of gifts requested by children sponsored by the "Angel Tree" program. They were to identify which of the bar graphs was the best representation of the answer to the reporter's question and to provide a justification.

On a second day, the students determined the costs to purchase 1, 2, 3, …, *n* gifts if each gift cost $5 or $10. Hence, students had an opportunity to generate a table of values, a general equation, and a graph of a linear function.

Notice that these activities covered number and operation, algebra, data analysis, problem solving, communication, reasoning, connections, and representation. How many typical textbook lessons integrate so many of the NCTM Standards?

## Books with Potential to Target the NCTM Standards

The books listed below are just a few that can be used in the middle school mathematics classroom to teach many of the topics emphasized in the NCTM Standards.

## Meaningful Learning through Literature

An old saying declares that "variety is the spice of life." If that aphorism is true in the real world, then it must be especially true in a mathematics classroom. The use of literature can be more than enrichment or a "fun" activity. Literature offers an opportunity to engage in interesting mathematics that integrates multiple strands in real-world situations. We encourage new and veteran teachers to *read in mathematics*. We guarantee that you and your students will have a great time!

## Small Groups in My Classroom?

*Art Johnson*

A number of studies have suggested that using small groups can improve students' learning in mathematics. *Principles and Standards for School Mathematics* (NCTM 2000) notes that small groups enable students to try out their ideas before presenting them to the entire class. Much has been written about the benefits of working in small groups—for example, students can learn from one another, and explorations can be more diverse. Much has also been written about the pitfalls—for example, assessment may be difficult, one student may do all the work, and students may not spend adequate time on task. Still, many questions remain, especially about the use of small

| Books | Number & Operation | Algebra | Geometry | Measurement | Data & Probability | Problem Solving | Reasoning & Proof | Communication | Connections | Representations |
|---|---|---|---|---|---|---|---|---|---|---|
| *The Gift of the Magi,* by O. Henry | • | • | | | • | • | • | • | • | • |
| *Can You Count to a Googol?* by Robert E. Wells | • | | | • | | | | • | • | • |
| *A Grain of Rice,* by Helena Clare Pittman | • | • | | • | • | • | | • | • | • |
| *The Greedy Triangle,* by Marilyn Burns | | | • | | | | | • | • | • |
| *Anno's Magic Seeds,* by Mitsumasa Anno | • | • | • | • | • | • | • | • | • | • |
| *The Librarian Who Measured the Earth,* by Kathryn Lasky | • | | • | • | | • | • | • | • | • |
| *Sir Cumference and the First Round Table,* by Cindy Neuschwander | | | • | | | • | | • | • | • |

groups in the upper grades. When should teachers use small groups? What types of topics or activities can be enhanced in a small-group setting? Given that high-stakes tests loom at the end of the year, how can small groups fit into a crowded syllabus? What is a good way to manage a classroom of small groups?

## Potential Small-Group Activities

The small-group setting can be an effective format to promote mathematics learning with many activities. The activities listed below are common to most mathematics classes and fit easily into a standard course syllabus. Using small groups for these activities does not require additional time, special topics, or extensive preparation. The suggestions below are a representative list of possibilities for using groups at all grade levels.

### Data gathering

Students often gather experimental data in class for analysis. When students work in small groups and pool their results, they are able to make conjectures on the basis of their data. Small-group data may also be combined to compare large-group results and to draw inferences. Some specific tasks that benefit from gathering and analyzing data in this way include explorations involving computer-based laboratory technology, investigations with interactive geometry software, or analysis of data from probability experiments, such as flipping coins. Students might also collect data outside the classroom, for example, from the Internet or student surveys. They might gather personal information, such as the time needed for each student in the class to travel from home to school or the size of each student's bedroom. This information can be compared or combined with the data of others in the group to make a conjecture or to construct a data display.

### Explorations

Students might work together in small groups to develop conjectures about how the graph of an equation changes as various parameters of the equation are altered. Small groups might explore the area and side lengths of similar figures to discover the relationship between linear and area measures. In both tasks, students can examine their findings in small groups to produce effective conjectures.

### Projects

Small-group projects can be beneficial activities for extending or deepening students' mathematics knowledge. Small groups of students can effectively report to the class about such topics as various types of geometries, consumer affairs, or real-world applications.

### Open-ended problems

Students can explore many open-ended problems in small-group settings. For example, the box problem requires students to determine the best way to cut out corner squares from a sheet of paper so that it will fold up to form the box with the largest volume. This problem, as well as those involving maximum-minimum points, promotes interesting discussions as students form conclusions on the basis of their results.

### Review sessions

Small-group settings can be effective for students who are reviewing material in preparation for major tests in the course or for standardized tests. Students work together to recall, clarify, and extend their knowledge of the course material.

### Group tests

Students may also be assessed in small-group settings. For example, students might be given a single copy of a test with the requirement that no erasures or changes may be made to answers once they are recorded on the test paper. Alternatively, small groups of students may be given a few minutes to discuss the entire test; then each group member takes the test individually. A third option is to distribute different portions of the test to individual group members, then combine their results for a single group score.

### Outdoor tasks

Students benefit more from outside activities or explorations when they work in small groups. For example, students might search the school for examples of the golden rectangle. A small group is also the best setting for using geometry skills to measure the height of a flagpole indirectly or to find the angle of elevation of the sun.

## Small-Group Management

After you have selected activities for small groups, you must consider small-group management. The logistics of setting up small groups can be problematic if left to chance.

### Group size

Small groups seem to function best if the group is composed of four members. This size is large enough to allow for meaningful discussion among group members but not so large that the group is unwieldy or has difficulty staying on task. Further, groups of four can easily be divided into pairs for some explorations.

### Group composition

The composition of a group is another important consideration. Groups in a classroom should be balanced according to ability and ethnicity. Early in the year, a balance of personalities is valuable, but this need should diminish in importance as the year progresses and students work in class together. Groups should be changed every two or three weeks to give students the opportunity to work with everyone in the class.

### Group-work goals

Establishing clear goals for small groups is essential. At the start of the year, you may need to assign specific roles to group members, such as reporter, recorder, materials manager, and so forth. Describing and even role-playing the various responsibilities of each group member may be helpful. Your assessment of small-group work might emphasize not only the final product of the group but also the involvement of each group member. That is, how well did each member contribute to the group effort? Was each member prepared, engaged, and active? Assessing the process, as well as the product, can help ensure that all members of the group offer valuable input to the group's tasks.

## Planning for Group Productivity

A productive group session does not happen simply by having students push their desks together to work on a problem. Successful groups require careful planning, from the activities and the group composition to the responsibilities of the students and appropriate assessments. But the result is worth the effort. All levels of students can benefit from participating in small groups as a regular feature of their mathematics classroom experience.

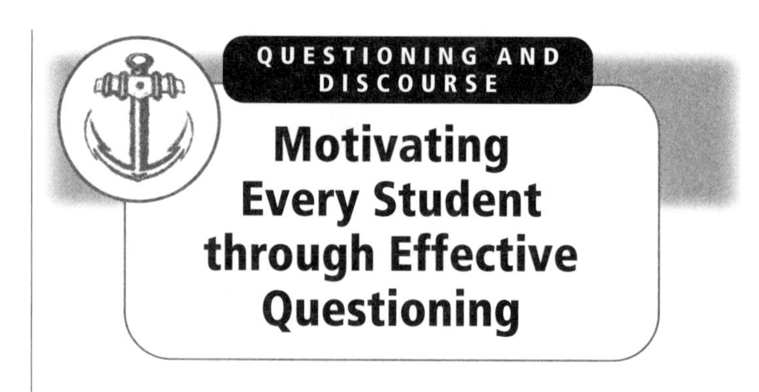


## Motivating Every Student through Effective Questioning

*Jane M. Wilburne*

Experienced teachers know that good, motivating questions can help keep students on task, prompt them to think about the material, and give them opportunities to reflect on their understanding of the lesson. Effective questioning engages all students, not just one or two in the class.

To be effective, questioning in the classroom should be used as both an instructional strategy and an assessment strategy. For students, the questions posed by the teacher should enhance their learning and encourage them to become involved in the lesson. For teachers, the questions posed should enable her or his assessment of the students' understanding of the material and the effectiveness of the lesson. If students appear confused and have no sense of how to answer the questions, the teacher should make the appropriate adjustments to be sure that each student understands.

Effective questioning should engage students in an exploration of the material. Questioning should require students to use critical-thinking skills, not simply to give rote answers. Students should be engaged in the learning and challenged by the questioning, and they should assume much of the responsibility for the discussions and explanations that take place in the classroom.

You should prepare high-quality questions at the same time that you plan your lessons. Giving some thought to the questions you will pose helps guide the lesson through an appropriate sequence of activities and keeps you and the students on task.

Keep the summary in table 1 (Wilburne) (p. 19) on your desk to remind you of various questioning techniques of effective teachers, but do not feel that you must master all these techniques in one lesson. Strive to incorporate one technique into your questioning each week. Effective questioning takes time and practice. Keep an index card on which you list various questioning techniques that seemed to work well and those that did not work well.

*(Continued on page 20)*

**TABLE 1 (WILBURNE).** ***Summary of Questioning Techniques***

| Ineffective Questioning Techniques | Effective Questioning Techniques |
|---|---|
| Asking yes/no questions | Asking questions that require short responses in which students must justify their answers, for example,<br>• "How did you know that answer?"<br>• "Can you explain how you found that solution?"<br>• "How can we figure this problem out?" |
| Calling out a student's name, then asking the question | Posing a question to the whole class, then pausing—for a long period of time if necessary. After the pause, try one of these tactics:<br>• Have students answer the question in their notebooks before you call on someone.<br>• Have students discuss their answers with partners before you call on one or two students.<br>• Inform students that you are going to call on two or three of them, but first, they must all think about the question. |
| Asking questions that are vague or misleading | Stating questions that are clear and target the learning goals |
| Answering your own question if no one in the class responds | Requiring students to work in pairs to discuss the question or write down their answers in their notebooks and share their solutions with partners |
| Asking, "Do you have any questions? Does everybody understand?" | Phrasing questions to determine whether the students understand, for example,<br>• "Who can ask the class a question about the lesson?"<br>• "Each of you should write down one question you have about the lesson. Then, ask your partner."<br>• "Did anyone come up with a question that his or her partner was unable to answer?" |
| Asking teacher-centered questions, such as, "Can someone explain to me…?" | Asking questions for the class, such as, "Can someone explain to us…?" |
| Asking questions that require the class as a whole to chant a response; for example, "Everyone, what kind of angle is this?" | Asking questions to help identify which students may not understand; for example, "Identify this type of angle in your notebook. When you are done, look up." |
| Asking only a few students questions throughout the period | Calling on as many students as possible |
| Informing a student that his or her answer is wrong | Guiding students through a series of questions to realize their errors |
| Asking the same type of questions over and over again | Asking a variety of questions in a variety of ways |
| Talking in a monotone voice | Getting excited and using inflection in your voice; whispering occasionally for effect like you have a big secret |
| Standing in the same location when asking questions | Moving around the room to make students aware that they are all involved in the lesson |
| Praising students who answer with a blunt "good" | Using pauses! After a student answers, ask other students what they think, for example, "Do you agree with his answer, Mary? What do you think, John?" |

*(Continued from page 18)*

Take time to reflect on how effective your questions were during your lessons. Did they motivate the students? Did they promote curiosity? Did they inspire the students to want to learn more? As you refine your questioning techniques, you will discover that you, too, learn from each question you pose. •

## A Short Story for Mathematics Teachers

*John E. Hammett III*

During the review of assigned homework problems, Jason volunteers his answer to the following question: "What is the area of a rectangular garden with dimensions 6 feet by 4 feet?" Jason says with some pride, "20 feet." His teacher begins to respond but then abruptly pauses.

Remembering some good advice about listening carefully to what students say and taking a deep breath, Jason's teacher asks him, "How did you get your answer?" The student explains that he doubled the length, resulting in 12; then he doubled the width, resulting in 8; he then added these two numbers to get the area.

Before saying that Jason is wrong, the teacher hesitates once again. An almost-imperceptible-yet-knowing smile forms on the teacher's face. "Jason, that's a correct answer to a good question, but it is not the question being asked in this problem. What question did you actually answer, Jason?" The student wrinkles his eyebrows in response to the teacher's comment and subsequent question, looking slightly puzzled. Opening the conversation to the rest of the class, the teacher asks, "What problem did Jason solve?" Melissa raises her hand; when the teacher calls on her, she says, "He found the distance around the rectangular garden. That's the perimeter."

The teacher thanks Melissa for her contribution and returns to Jason to verify his understanding. Jason nods in agreement. He takes a quick look at the diagram of the rectangular garden that he drew in his notebook and asks, "Is the correct answer 24 square feet?" The teacher again asks him to explain his answer, and he does. This time, Jason says that he multiplied the rectangle's two dimensions to find the area of the garden. When the teacher asks, "Is this the correct answer to this question?" Jason replies with confidence, "Yes!"

As the class discussion expands to include additional homework problems and other students, the teacher realizes that this incident should have a positive impact on how Jason learns mathematics. By listening to the student's solution, the teacher, as well as Jason's classmate, Melissa, helped him correct his own misconception, thereby fostering success and self-confidence instead of failure and self-doubt. And Jason, Melissa, and the other students in the class competently and confidently calculated rectangular areas happily ever after.

## Using Learning Logs in the Mathematics Classroom

*Roni Jo Draper*
*Margaret E. McIntosh*

Learning logs provide space for students to respond to writing prompts. Our use of learning logs has allowed us to know our students better, to understand their thinking better, to communicate individually with students through the written word, and to reevaluate our instruction on the basis of students' responses. The types and uses of learning logs can vary widely; what follows is a partial list of applications for learning logs in the mathematics classroom.

- Learning logs can be used to open a lesson, readying students for the topic and allowing the teacher to assess students' knowledge of material that is to be presented. Sample prompts might include "What do you already know about slope [or another concept]?" "What do you think you might learn today about slope?" "What do you need to learn about slope?"
- Learning logs can also be used to conclude a lesson, helping students reflect on what they have learned and identify gaps in their understanding after instruction. Sample prompts might include "What did you learn about slope [or another concept] today?" "What questions do you still have about slope after today's lesson?" "How does what you learned today about slope fit with what you already knew about slope?"
- Learning logs compel students to articulate their thinking. The following writing prompts can shed light on

how well students understand mathematical concepts: "Choose the hardest problem from today's assignment, and explain how you solved it." "Find the error in the following problem, and explain how to solve this problem without an error." "Solve the following problem in two different ways, and explain why both ways work."

- The following prompts can help teachers evaluate their students' attitudes and biases: "How do you feel about mathematics?" "Why are you taking this class?" "Describe the ideal mathematics class."
- Finally, these prompts can help students reflect on their study strategies and skills: "What do you do when you get stuck on a homework problem?" "How do you take notes for this class, and what do you do with the notes after class?" "How do you prepare for tests and quizzes?"

Students can use about five minutes at the beginning or end of class to respond to prompts, thereby allowing the teacher to read and reply to students' writing quickly. Here are some more hints for using learning logs in the mathematics classroom:

- Use learning logs frequently, at least several times a week.
- Do not accept partial, ill-conceived, or no-effort answers; simply have students rewrite their responses until they have met the standard.
- Respond to students' writing, even briefly, to make sure students know that their learning logs are being read. •

## Problem Posing: What's in a Word?

*Fiona Thangata*

Give students opportunities to write mathematics problems. Solving a problem may involve only applying a recently learned technique, but writing a similar problem requires a deeper understanding of the underlying mathematical structure of the problem. The contexts that students choose for embedding their problems also reflect their interests, concerns, and background knowledge and can motivate students to connect mathematics with their daily lives.

## Teaching Students to Think Well: Ideas for Beginning Teachers, K–8

*Cheryl Franklin*

As a beginning teacher, I was concerned about the thinking and learning of my students. To meet their diverse needs, I used a variety of curriculum and instructional materials. My students seemed to thrive. However, I kept wondering, "How do I help my students *think* well?" Although I was doing "all the right things" and using wonderful materials, I did not know whether I was helping my students become powerful mathematical thinkers or whether I was simply providing them with entertaining activities. Since nifty activities are not enough by themselves, we need to consider ways to implement such activities so that students learn to think well in mathematics.

### Teaching Strategies Based on Cognitive Science

My concern about fostering students' thinking abilities led me to the field of cognitive science, the scientific study of the mind. Several teaching strategies have been based on findings from researchers who study how the mind processes information. What I have learned has improved my teaching and given me a foundation on which to base my professional decisions. Three major strategies are relevant for the beginning teacher.

#### Ideas from process tracing

One strategy comes from *process tracing*. Simply put, process tracing is having people think aloud as a way of learning what they are thinking. The focus is on the child's mathematical thinking. Process tracing in a classroom can begin with something as simple as interviewing students individually. The interview need not be lengthy—just framed with questions that target what the student is thinking (e.g., "How did you solve this problem?" "Tell me what you thought of doing?" "Can you explain why you chose this strategy?"). Given the time constraints in classrooms, it may be easier to interview a sampling of students. What you learn from this sampling can enlighten you on the thinking of the entire class. You are then placed in a better position to make curriculum and assessment decisions for your students.

### Ideas from research on conceptual change and misconceptions

Other strategies to help students think well come from research on conceptual change and misconceptions. Researchers and teachers alike know that students can hold many misconceptions in mathematics. For example, students might think that subtraction of two-digit numbers is simply "taking the smaller from the bigger." Subtraction is taught in the primary grades, yet in the upper grades a simple problem, such as 47–29, commonly results in an erroneous answer, such as 22. In an effort to correct this misconception, many teachers would suggest providing manipulatives to help students better understand *regrouping*. Using concrete models may help students correct their misconceptions, but using manipulatives alone does not ensure understanding. Students' activity with manipulatives depends on our guidance as teachers. If we are to promote mathematical understanding, we must help our students clearly make connections between manipulatives and mathematical ideas. Otherwise, misconceptions will remain prevalent in students' thinking and we will have fallen short in our attempts to help them think well.

### Ideas from research on expertise

Other strategies flow from research on expertise. The discussion has been focused on comparing experts and novices. Experts understand the underlying structure of content; novices represent the surface appearance of problems. As educators, we grapple with the questions of "How do we educate for expertise?" and "What kinds of experiences can we provide for students that will lay a foundation for the development of expertise?"

One potential strategy can be used as we plan our curriculum or assess the work that students do. For example, in a classroom in which data analysis is the content, a teacher might ask, "What do we do in real life with data?" and build the content from that question. Students can collect, represent, and interpret real data. Although their work may differ in many ways from that of adult expert statisticians, their processes are similar. Students can pose a question that they are interested in investigating, develop a sampling plan, collect data, analyze it, and use the data to describe and make decisions about real situations. In a pragmatic way, students are then learning to think well. Although they may be novices in that area, we can help them think well by having them be more explicit in their thinking about the underlying mathematical concepts they are encountering.

## Conclusion

The field of cognitive science need not overwhelm the beginning teacher. The three areas of process tracing, conceptual change, and research on expertise have practical applications to the mathematics classroom that can influence our teaching and help us think about the possibilities of helping our students think well. •

## Say What You Mean

A favorite story at our school involves teaching decimals. A student was given a decimal addition question in a horizontal fashion, 6.9 + 3.82 + 4.257 = *n*. He told the teacher that he could not do it, so she responded by telling him to "rewrite it vertically and line up the decimals." This is what the student did:

6<br>.<br>9<br>3<br>.<br>8<br>2<br>4<br>.<br>2<br>5<br>7<br>+<br>—

From this experience, we learned to be a little more precise in our directions and not assume that students always understand what we mean.

—*Kim McLean*

## Something I Never Learned in Methods Class: Know Your Students

Do not expect every student to be as excited about mathematics as you are. Some students will share your enthusiasm, but many will not. Get to know what outside interests your students have. Find out what things they excel at, and celebrate those accomplishments with them. Consider how these areas could be used in contexts for mathematics.

—*Margaret R. Meyer*

TOOLS

# Mathematics Learning with Technology

*Ed Dickey*
*Melina Deligiannidou*
*Ashley Lanning*

The purpose of using technology is not to make the learning of mathematics easier, but richer and better.

—*Alfinio Flores*

The first axiom in technology planning is, Let the mathematics drive the lesson. Look for ideas that allow you to use technology to enhance your students' understanding of the mathematics they are learning. Avoid teaching technology for the sake of technology. Try to attend an NCTM or a state mathematics conference; both abound with sessions in which experienced teachers explain and demonstrate how they use technology with their students.

Remember to start small but to *start*. If you incorporate technology just once a semester in each class you teach, you are making progress. By adding one more technology lessons each semester, you can, within five to ten years, become an exemplary technology-using teacher.

Students often need orientation to the activity you have planned. Realize that once they are working with the computer, they will no longer be paying attention to you. Technology removes the locus of control from the teacher to the computer, calculator, or group of students using the technology. Plan your lesson in phases. You might begin with an introduction to the whole class. Explain the specific task for students to accomplish or the problem you wish to have them solve. For some activities, you may be wise to pose the problem in vague terms to allow your students a wide range of options, but keep in mind that some students need more structure or direction. A well-constructed worksheet with specific questions can scaffold an otherwise unproductive investigation.

Your role with technology is that of a coach. Resist having students work individually. Working in groups minimizes the number of questions you have to answer and allows students to help one another. Provide opportunities for students to share their work with the whole class. This activity allows you to view their presentations from the sidelines and think about the mathematics instead of what you plan to say next.

Using technology as part of student assessment is a matter of consistency; after all, if you believe that technology enhances learning, then it will also augment assessment. Further, using technology increases the authenticity of assessment, making it more relevant to the world of the student. Performance tasks offer an effective method for using technology in assessment.

Regardless of what software you use, students need time and assistance to learn the application. Often, the more powerful the software, the more effort required to learn to use it. Spreadsheet, interactive geometry, and computer algebra system (CAS) software require students to apply themselves to work effectively. Calculators require orientation. Invest class time to ensure that your students develop useful technology skills. This investment will pay off with enhanced learning of both the technology and the mathematics. •

## Hands-Off Technology Demonstration

The best strategy for demonstrating technology may be a "hands-off" approach. Do not touch the hardware when demonstrating to your students how to use, for example, a spreadsheet or a graphing calculator. Instead, select a student who has limited experience using the technology that is to be demonstrated; have the student follow your directions while using a workstation that is projected to the entire class. You are then free to move about the room while giving directions and monitoring your students' progress. This "hands-off" approach regulates the pace of the presentation, allowing students to keep up with the demonstration and focus on the topic under discussion.

—*Todd Johnson*

Todd Johnson passed away after this tip was accepted for publication.

# Maximizing Manipulatives

*Thomasenia Lott Adams*

The purpose of a manipulative is to help students learn the mathematics. To ascertain whether a manipulative meets this goal I use three steps. First, I explore the manipulative *before* instruction to determine the best ways to use it and to discover any hidden "kinks." Second, I observe students using the manipulative *during* instruction to determine how well these students develop an understanding of a specific topic while using that manipulative. Finally, *after* instruction on the basis of my observations, I assess whether the manipulative met the goal that I set for it prior to instruction. My ultimate goal is to find out whether the manipulative served its purpose to support and enhance students' mathematics learning.

I have developed a grid to help me keep track of the manipulatives that I have available and the concepts for which I have found them useful. When I obtain a new manipulative, I add it to the list. The grid is a flexible tool that can be modified as needed for individual classes or groups of students, and it can be used to choose manipulatives to lend variety, reinforcement, and enhancement in mathematics instruction. Table 1 (Adams) shows my grid for maximizing manipulative use.

**TABLE 1 (ADAMS).** ***Maximizing Manipulatives***

| Manipulatives | Number Sense | Whole Numbers | Fractions | Decimals | Operations | Algebra | Geometry | Measurement | Data Analysis | Probability |
|---|---|---|---|---|---|---|---|---|---|---|
| Attribute blocks *† | ● | | ● | | | | | | | ● |
| Base-ten blocks *† | ● | | ● | ● | ● | | | | | |
| Plain/color cubes | ● | ● | ● | | ● | ● | ● | ● | | ● |
| Color tiles * | ● | ● | ● | | ● | | ● | ● | | ● |
| Counting chips *† | ● | ● | | | ● | | | ● | | ● |
| Cuisenaire rods *† | | | ● | | ● | ● | | ● | | |
| Dot/number dice | ● | ● | ● | | ● | | | | | ● |
| Dominoes * | ● | ● | ● | | ● | | | | | |
| Fraction circles *† | | | ● | | | | ● | | | |
| Fraction rods *† | | | ● | | | | | ● | | |
| Fraction squares * | | | ● | | | | ● | | | |
| Geoboards *† | | | ● | | | | ● | ● | | |
| Geometry solids † | | | | | | | ● | | | |
| Graph paper *† | ● | ● | ● | ● | ● | ● | ● | ● | ● | |
| Links *† | ● | ● | ● | | ● | ● | | ● | ● | ● |
| Mirrors † | | | | | | | ● | | | |
| Pattern Blocks *† | ● | | ● | | | | ● | | | |
| Patty Paper † | | | | | | | ● | | | |
| Playing cards * | | | | | | | | | | ● |
| Spinners *† | | | | | ● | | | | | ● |
| String † | | | | | | | | ● | | |
| Tangrams *† | | | ● | | | | ● | | | |
| 2-Color counters † | ● | ● | ● | | ● | ● | | | | ● |
| Unifix cubes † | ● | ● | ● | | ● | | ● | ● | | ● |
| Weights | | | ● | | | | | ● | | |

* Transparent form is available for whole class demonstration.

† Classroom set is available for small group or individual exploration.

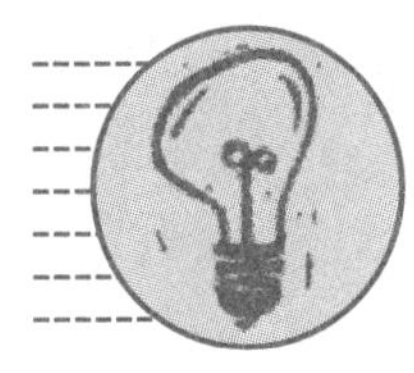

# Using the Hands-on Approach

*Gail Englert*

I make sure that manipulatives and calculators are visible in the classroom and are easy for students to access. I allow my students to explore these materials when they have free time. Then, when the opportunity arises to use the items to learn about mathematics, the novelty has worn off, students are familiar with them, and the materials serve the purpose intended. When solving problems, I allow my students to select the tools that they think will help them. By watching what materials my students select and how they use them, I get glimpses into their mathematical thinking, including their problem-solving abilities. Sometimes I even learn new ways to consider a problem!

3 SECTION

# CLASSROOM ASSESSMENT

As a beginning teacher, you have many concerns that you need to address every day. One of your most important concerns is how to determine what your students are learning as a result of their experiences in your classroom. Classroom assessment encompasses much more than simply a test at the end of a unit; it includes every action you take to determine what your students know and understand during every lesson you teach. *Principles and Standards for School Mathematics* (NCTM 2000) states that assessment "should support the learning of important mathematics and furnish useful information to both teachers and students" (p. 11). Assessment should be a component of all mathematics lessons to inform you about your students' learning and guide your instructional decisions.

## Uses of Classroom Assessment

Classroom assessment can be a useful tool for both you and your students. Your students can use the information from assessments to set goals and to gauge their progress toward those goals. You can use the information to help make decisions about your teaching. Gathering evidence of students' learning helps you identify those who need additional support or challenges and determine your next instructional steps.

## Types of Assessment

Assessments come in a variety of forms, including learning logs, journals, observations, interviews, student self-assessments, performance tasks, projects, portfolios, and more traditional tests. Each assessment technique gives you a different way of looking at your students' understanding; thus, choosing multiple forms of assessment is important in developing a broad picture of what your students know and can do. For example, students can make a presentation to the class about a problem they solved, assess their own use of problem-solving strategies while working on the problem, and select a solution for inclusion in their portfolios.

## Planning for Assessment

An essential first step in planning assessments is to identify the mathematics you want students to learn and the evidence you need to gather to determine the extent of their learning. As you plan your lesson, think about what you want your students to learn and how you will know that they have learned it. Assessment should be woven into your lesson as a part of the learning experience. To guide your planning, ask students to write a "K-W-L essay," describing what they ***know***, what they ***want*** to know, and what they ***learned***, as you begin and end a unit. Deciding what to assess, what assessment to use, and what you will do with the information gathered from assessment should all be parts of your lesson planning.

## Grading Assessments

Grading provides the data you need to use assessments effectively. Different forms of assessment lend themselves to different methods of grading. Sometimes, simply reading and responding to your students' learning logs is sufficient. At other times, you and your students will need to gather more extensive data about their learning. Many assessments can be graded using rubrics. If you choose to use a rubric, consider having your students help you develop the grading criteria before the assessment is given, to clarify your expectations. Be selective in your assessments and how you grade them, however, to ensure that you and your students are not overwhelmed with papers!

As you become more proficient in incorporating assessment into your lessons, gathering and interpreting data from assessment, and using that information to guide your teaching practice, your students will learn more and planning a lesson will become easier and more efficient.

Obviously, classroom assessment is difficult to separate from planning. As a result, you will find assessment ideas throughout this book, not just in this section. In particular, look for ideas in the "Curriculum and Instruction" and "Classroom Management and Organization" sections. As you read this section and think about how you will evaluate your students' understanding, you might keep the following questions in mind:

- What is the important mathematics I want my students to learn, and how will I know that they have learned it?
- What types of assessment do I want to use to gather evidence of student learning, and when is each most appropriate?
- How can my students and I use the data I collect from assessment to improve their learning?

Photograph by Makoto Yoshida. 

# Strategies for Tapping "Hidden" Learners

*Mary C. Shafer*

As a high school teacher, I try to shape my daily instruction by taking into consideration my students' understanding, insights, and misunderstandings, but I often wonder what those students who rarely engage in classroom conversation know about the mathematics they are learning. I have found the following four methods to be especially effective for opening the lines of communication.

## Pair Work

Often, students are more willing to talk when they work in pairs. During pair work, I listen for students' approaches to problems, their understanding of the mathematics, and the mathematical vocabulary they use. I ask students who use unique approaches to explain them in class discussions. In this way, all students learn to see mathematics as a creative process that is within their capabilities.

## Journal Writing

Give students opportunities to write in their journals for five to ten minutes once or twice a week. Most students take this assignment seriously because it gives them the opportunity to talk with me directly without other students' becoming aware of their concerns. Students who are shy or popular or whose cultural backgrounds may limit participation in whole-class discussion use the journals as a way to communicate specific needs, difficulties, or successes in understanding concepts. The brief writing tasks provide clues about what students are learning. Over time, I have learned the types of questions that effectively tap into their thinking.

## Student Presentations

Some problem solutions are especially good for students to share orally in class, for example, problems set in various contexts that involve application of mathematical concepts. On a given day, each group of three to four students presents a solution to one of several assigned problems. Group presentations provide a forum for acceptable risk taking. Students present the group's solution, not their own, and all members of the group answer questions if the presenter is unsure of a particular aspect. Over time, every student in the class has the opportunity to present a group solution.

## Projects

Projects offer opportunities for students to think about mathematics in different ways or to explore applications of mathematics in everyday situations. Every grading period, I ask students to complete one project from a set of eight options that are directly related to the mathematics studied during that grading period. Students may also propose their own ideas for these projects for teacher approval. I display all satisfactory projects in the classroom or school showcases. Students often talk about the projects with other students before and after class. Some students are excited about their projects, ask to talk about them in class, and share their projects with teachers in other disciplines.

## Learning Discoveries

Some students in my classes are active talkers, but many are quiet—my "hidden" learners. Through listening to discussions with partners and reading their journals, I continually learn about the content they understand, the types of problems that cause them difficulty, and the misconceptions that they may have developed. I use this information to try different approaches in presenting content and to focus whole-class discussions. In group presentations and projects, students show their peers and me what they are learning and are excited about. In the end, I learn more not only about my students but also about myself and my strengths and weaknesses in teaching them. Using these methods, we all learn and relearn mathematics—and we are enthusiastic about the process.

## Let *All* Students Show What They Know

**To allow *all* students to show what they know, in your mathematics instruction, vary their assessments to include oral and written responses and long-term projects. Clearly communicate your expectations for each assignment in advance.**

*—Patricia McCue and Jennifer Lana-Etzel*

## Student-Developed Rubric: A Work in Progress

*Ranjani Sriram*

How can students identify their strengths and weaknesses from a scored assessment and learn from that information? This nagging question led me to create rubrics for scoring certain types of assessments. Specifically, I involved my students in creating a rubric for problem solving. This activity helped them gain an understanding of my expectations for solutions to multistep problems.

Students were assigned to work in groups of three or four. At the start of class, we conducted a brainstorming session to answer the following question: What are the important components of a solution to a multistep problem? We then allocated the ideas that the students offered into five categories that I consider important to assessment: (1) organization, (2) communication, (3) strategy or procedure, (4) accuracy, and (5) completeness. In some classes, I modified this process, listing the five categories first, then having students identify the components of each category.

Each student group was assigned one category. On the basis of the requirements, students used their own words to develop a rubric. That is, for the given category, they had to describe what kind of work would earn ratings of 4, 3, 2, and 1, with 4 as the highest rating. Students displayed their ideas on large pieces of paper. I allotted an entire forty-five-minute period for this activity and conducted it in all five of my classes. At the end of the day, I compiled the information from all classes and made one rubric for all the categories (see fig. 1 [Sriram].

The next day, every student was given a copy of the rubric that they all had created. I told them that I would use this rubric every time I gave problem-solving assessments, and I encouraged them to refer to it when they completed the assessments. Student pairs were then given a few anonymous sample solutions to a word problem that other students had previously completed; the pairs worked to score each of these student samples using the new rubric.

This activity was beneficial for my classes and encouraged student involvement. When students scored the sample problems, I heard such comments as "This is a great way of doing the problem; I never thought of that"; "This step was really not necessary"; "I never thought grading was so much work"; "The work shown makes sense; I get it now"; and "How did they come up with that number?" These comments told me that the students were studying each step carefully and thinking deeply about the process. A follow-up reflection sheet revealed that students valued the rubric-creating activity because the result was written in their language and was easy to understand.

I enlarged the rubric to poster size and had it laminated. Subsequently, when I gave students a problem-solving quiz, they could refer to the rubric during the quiz. They seemed to put a good deal of effort into their work to achieve higher ratings. This activity capitalized on student accountability, responsibility, pride, and involvement—all of which need frequent reinforcement.

**FIGURE 1 (SRIRAM)**
***Rubric for problem solving***

**Organization**

| | |
|---|---|
| 4 | All steps are neatly shown. The necessary information is categorized in charts, tables, or lists. All written work is legible, and if necessary, it is in paragraph form. |
| 3 | Most work is neatly shown. Charts, tables, or lists are neat, with most of the necessary information organized. Most of the written work is legible. |
| 2 | Half the work is missing. Charts, tables, or lists are incomplete. Written work is not too legible. |
| 1 | Steps are very unclear. The necessary information has not been organized at all. Written work is illegible. |

**Communication**

| | |
|---|---|
| 4 | Explanation of steps is very clear and thorough. All work shows a clear understanding of the problem. |
| 3 | Explanation of steps is good but could use more detail. Most work shows an understanding of the problem. |
| 2 | Explanation of steps is poor with very little detail. Work shows little understanding of the problem. |

(Continued on next page)

FIGURE 1 (SIRAM) (continued)

| | |
|---|---|
| 1 | Very poor explanation with no understanding of how to solve the problem. Work shows no understanding of the problem. |
| **Strategy/Procedure** | |
| 4 | Method and steps used to solve the problem are clearly displayed. Strategy used shows a clear understanding of the problem. |
| 3 | Most of the steps used to solve the problem are clearly shown. Strategy used shows a somewhat clear understanding of the problem. |
| 2 | Few steps are shown in solving the problem. Strategy does not show an understanding of the problem. |
| 1 | Barely any steps shown. Strategy, if any, is random. |
| **Accuracy** | |
| 4 | All calculations are done correctly. All information from the problem is accurately interpreted. Checked to see if answer makes sense. |
| 3 | Most calculations are done correctly. Most information from the problem is accurately interpreted. Checked to see if answer makes sense. |
| 2 | Half the calculations are done correctly. Information from the problem has not been clearly interpreted. Did not check to see if answer made complete sense. |
| 1 | The answer is incorrect. Information from problem is misinterpreted. Answer does not make sense. |
| **Completeness** | |
| 4 | All work is complete. All directions are followed. All questions are answered. |
| 3 | Most work is complete. Most directions are followed. Most questions are answered. |
| 2 | Half the work is done. Very few directions are followed. Half the questions are answered. |
| 1 | Work shows minimal to no effort. |

## Perspectives on Scoring

*John E. Hammett III*

How can you encourage your mathematics students to value the process of solving a problem as much as the product of obtaining an appropriate answer to a question? First, avoid answer columns on homework assignments and tests, all-or-nothing grading schemes, and the tendency to label student work as wrong when it is not entirely correct or when it uses a method that differs from yours. Instead, insist on fully detailed mathematical explanations whenever your students solve problems, rewarding every reasonable effort—even those that use a different approach—with positively oriented credit accumulation rather than negatively oriented credit reduction. In other words, use the notation "+2 out of 4 points" instead of "–2 out of 4 points," especially when the solution is incomplete or only partially correct.

## Evaluating Progress

Be aware that tests should measure what students know, not what they do not know. Give tests that assess whether students know basic concepts and that allow you to differentiate among their levels of understanding. If you cannot complete your test in a quarter of the time that your students will have to take it, then the students will be unable to complete the test in the time allowed.

—*Agnes M. Rash*

# Using Notes during Tests

*Todd Johnson*

Before the first test in a mathematics class, students sometimes ask, "Can we use notes on the test?" In my class, students are often allowed to use notes when taking tests, but I limit the notes a student may use to one page. Additionally, students' notes must be submitted at the beginning of class the day before the test.

Allowing students to use notes during tests has several benefits, including the following:

- Students must review and synthesize their class notes to make a useful page of test notes.
- Because they must submit their notes the day before the test, students spend at least two nights studying.
- While making their pages of test notes, students come up with good questions to ask in class the day before the test.
- After collecting students' notes, the teacher can pinpoint concepts that students think are important to know for the test, identify students who might have difficulty learning particular concepts, and detect errors in students' thinking.
- Finally, students may become aware that tests do not always emphasize the recall of memorized information but often emphasize the application of that information.

---

Todd Johnson passed away after the writing of this item was completed. The editors have chosen to leave the text in the present tense, as it was prepared.

# Using Collaborative Testing as an Alternative Form of Assessment

*Joan Kwako*

One effective way to both gain understanding of students' knowledge and to provide students the opportunity to learn is through collaborative assessment. Collaboration requires students to defend their positions and explain their thinking, which, during a high-stakes situation such as a test, can enhance their learning because of the increased motivation. Collaborative-group tests require students to communicate and participate in groups. However, before you simply assign students randomly into groups and hand them a test, you should make some provisions for the composition of the groups and the type of test administered.

## Collaborative Group Size and Heterogeneity

Two factors to consider when assigning groups are size and heterogeneity. I have found that groups of three work best. In a pair, one student may easily dominate the assignment, and in groups of four, some students may get lost in the discussion. Making sure that groups are heterogeneous in terms of gender, status, and ability greatly increases their effectiveness. To ensure heterogeneity of all three factors may seem impossible—and sometimes, it is. Most likely, you will not be able to form perfectly heterogeneous groups, but keeping group composition in mind is important when asking students to complete group tests.

## Types of Collaborative Tests

The tests that are administered to groups must be designed differently from those that are given to individual students. Designing questions for a group test is more like creating open-ended homework problems than writing traditional test questions. Collaborative-test problems must require input from all group members; they should not be problems that can be solved by one person working alone. The questions must elicit discussion and encourage students to invest their effort in the solutions. Through these discussions, students can develop new ideas of their own on the basis of other students' thinking, expose common misconceptions, clarify their thinking, and justify their positions before reaching a group solution.

Another advantage is that this kind of assessment requires you to grade only one-third of the usual number of tests. Competitive students may be unhappy, however, that they are forced to depend on the work of others and may require some guidance in working collaboratively.

### The single-solution technique

One possible collaborative assessment technique has all group members solve one problem together and write a single solution. Students then individually answer questions about the group solution and solve two similar problems, one that is parallel to the group problem and one that extends it (Kroll, Masingila, and Mau 1996). This approach enables the teacher to calculate both an individual score and a group score. In a variation of this assessment, each student's grade is based on the sum of the group score and the average of the individual scores, resulting in the same grade for each group member. Again, this approach encourages students to commit to helping each of the group members learn the material, because each group member's score depends on the learning of everyone else in the group.

Another variation on this kind of grading includes awarding additional points if all group members score above a certain level on the individual portion of the test. This approach changes the dependence on others for each student's score to a commitment to the learning of all group members, creating a "one for all and all for one" atmosphere. One advantage of this approach is that students have the benefit of learning while taking the group portion of the test but maintain a level of individual accountability. One disadvantage is that fewer concepts can be tested because the three questions for each topic must be parallel in structure. Of course, for the teacher, this approach doubles the number of tests and, thus, the time necessary for testing and grading.

### The group take-home examination

Yet another approach to collaborative testing is the group take-home examination. Group members are given seven to ten days to complete the examination and hand in one solution signed by all, and all receive the same grade. In these situations, having group members anonymously grade all other members, including themselves, helps to determine whether everyone contributed equally.

A variation on this technique is the group oral take-home examination (Crannell 1999). In this situation, students are graded entirely on their understanding of the material as they present it; the written work that is handed in is used only to clarify points in their presentations. Students learn from their mistakes because "they get feedback even as they present their results, and since they have debated the results with their teammates, they care about the answer" (Crannell 1999, p. 144). The longer time allowed for this kind of assessment enables high-level conceptual problems to be included on the test, and students can learn a great deal when researching the answers to such problems. Teachers may have difficulty, however, determining whether all group members contributed equally and what each group member knows individually.

## Benefits and Drawbacks of Collaborative Testing

When administering group tests, keep in mind that students have probably not experienced similar assessments in the past. Some will revel in the idea; others will find it frustrating. Time spent on discussing effective group interactions may be helpful before implementing group tests.

Collaborative tests have many benefits: they are more realistic in terms of what students will face in the world of work in the future; they reduce the anxiety surrounding testing; they mimic more closely the actions of mathematicians; they increase students' potential to succeed; and most important, they afford students the opportunity to learn.

## Conclusion

This list of ideas for group testing is short. However, it may be enough to encourage you to consider assessment strategies other than the traditional lineup of homework, quizzes, and tests and to think of assessment as an opportunity for students to learn cooperatively.

Notes:

4 SECTION

# CLASSROOM MANAGEMENT AND ORGANIZATION

Experienced teachers tend to use a variety of methods to manage and organize their classrooms. In contrast, beginning teachers may wander the halls during the days before the start of school, peering into the classrooms of experienced teachers in search of ideas about how to arrange the room for that first day, how to set expectations that keep students on task, and how to keep track of students' work when they do stay on task! As a new teacher, the resources you find to address these concerns will influence your own strategies for management and organization and help you create a pleasant and productive learning environment. During your beginning years of teaching, among your first priorities should be the need to organize your classroom and manage the instructional environment in ways that support, rather than impede, students' learning of mathematics. The following paragraphs highlight some important aspects of classroom management and organization.

## Set Clear Expectations

Of course, your students are required to abide by your institution's rules and regulations, but you should also establish expectations regarding how your students should treat one another in the classroom and how they should accomplish their work. Do not hesitate to inform your students of these expectations and the consequences of not living up to them. Be fair and consistent in holding students to these expectations.

## Know One Another

Whether you are with your students for the entire school day or for only a single class period, arrange opportunities to converse with them. In the same manner that you are different from every other teacher, each student is also unique. One of the joys of teaching is getting to know each student as an individual. You should also provide opportunities for students to interact with one another. Building a sense of community is one important way to create an effective instructional environment in which students can learn mathematics.

## Organize People, Space, and Materials

Try a variety of arrangements for individual and group work to meet students' differing needs and keep them engaged in learning. As you get to know your students, you will be able to select appropriate arrangements for different tasks. Keep in mind, however, that students also need chances to try new methods. For instance, a student who seems to work best alone might be asked to work with a partner on a more complex task to help the "loner" become more comfortable working in a group.

You may also try different strategies to organize your space and materials. Experienced teachers have many wonderful ideas for arranging classrooms, making efficient use of closets, or setting up filing systems. Be on the lookout for new ideas, but always consider them in light of your style and your students' needs. Remember that the goal is for students to learn mathematics; your methods for organizing people, space, and materials should always contribute to—not detract from—your students' learning opportunities.

## Manage Time

You will soon learn that in teaching, you never have enough time! As you gain experience, you will develop ways to save more time for instruction. Learn to build natural "breaking points" into lessons that are difficult to teach as a whole in the time allowed. As you become more familiar with the required record keeping, look for ways to reduce time spent on that task, perhaps involving students or parent volunteers. Every few weeks, evaluate one hour of your instructional time to see whether you could make better use of the time by changing some small aspect of classroom management or organization.

## Assess Strategies for Organization

Not all management tools will work for all teachers. As you read this section and experiment with different strategies for classroom management and organization, you might evaluate each strategy by asking the following questions:

- How does this management or organizational idea contribute to a classroom environment that supports students' learning of mathematics?
- How does this idea help me better organize my students, space, or materials so that I can use instructional time more efficiently?
- How can I involve students in developing ideas for classroom management and organization to build their sense of ownership in the instructional environment?

Photograph by Makoto Yoshid. 

# REMINDERS FOR BEGINNING TEACHERS

*Chandra Guckeen*

The ideas listed below are reminders to help you put a classroom-management plan in place. Implementing these ideas at the beginning of the school year can prove beneficial throughout the entire year.

1. ***Try to greet every student at the door.*** Each child should be acknowledged every day with a "hello" and a handshake. I have found this habit to be effective in building relationships with students.

2. ***Make a seating chart.*** This aid helps you learn all students' names as quickly as possible. During the first few days, use the chart to quiz yourself at the end of each period. Calling each student by his or her name is very important in establishing rapport.

3. ***Get to know each student as a person.*** Students want to know that you are interested in them as people rather than just as students. To show your interest, host a club, coach a sport, or attend events and activities in which students participate. Doing so gives students an opportunity to see you in roles other than that of "teacher."

4. ***Establish rules, and consistently enforce them.*** Your students should know what the classroom rules and consequences for infractions are, so that no surprises arise when it comes to discipline. Be consistent, firm, and friendly, and maintain a sense of humor.

5. ***Post entering and exiting procedures.*** Students should know what is expected of them when they enter and exit your classroom. Establishing routines creates a consistency that middle school students need.

6. ***Create classroom jobs.*** These jobs could be Homework Collector (collects homework from each group), Paper Distributor (passes out papers during class), Return-Papers Person (passes back graded papers first thing when students enter the room), and the Drawing Supervisor (pulls the name out of the jar for the weekly prize drawing).

7. ***Have a weekly prize drawing.*** Give students tickets for positively contributing to the classroom in some manner. Students simply write their names on the tickets and then place them in a jar for the weekly drawing.

I hope that these reminders have given you direction in the task of planning your classroom management for the year. Remember, the overall goal in classroom management is to maintain a safe and effective learning environment for students.

# Helpful Hints for Teaching Active Youth

*Pamela Moses-Snipes*

As you embark on your first year of teaching and begin to develop your classroom-management skills, consider the following helpful hints to assist you in working with learners who may be more active than necessary and require stronger managing skills.

✓ ***Hint 1:*** Create engaging mathematics activities. Connect the mathematics concepts in the activity with mathematics that students use in their everyday lives.

✓ ***Hint 2:*** Create extensions of your activities to challenge your students. Extensions could be in the form of special projects, interdisciplinary connections, multicultural connections, or connections between mathematics and literature.

✓ ***Hint 3:*** Consider assigning several tasks within an activity or assignment to change the pace. Allow designated times for completion of the tasks. Having students complete several shorter tasks within a longer activity may minimize off-task behavior and thus help students remain focused on, and engaged in, completing each task before moving to the next task.

✓ ***Hint 4:*** Always enforce your classroom rules. If a consequence for students' actions has been established, be sure to impose it. Do not allow your classroom-rules list to be a mere decoration on the wall.

# Organizing for a Classful

*Cindie Heinrich Donahue*

I have been organizationally challenged my entire life. When I first became a teacher, however, I knew that I needed to change for my students' sake. Over the years, I have made great strides in organization. The following paragraphs highlight some wonderful ideas that have helped me the most.

## Organize Assignments

One idea involves a way to organize assignments for students who are absent. Although setting up this system initially took time, I have found that the effort was well worth the result. Each assignment is given a number. First-quarter assignments begin with an *a*; thus, the first assignment that students receive is numbered a-1, followed by a-2, and so on. Second-quarter assignments begin with a *b*, third-quarter assignments begin with a *c*, and fourth-quarter assignments begin with a *d*. This notation also enables me to write a shorter assignment label in my grade book. I reserve space on my bulletin board to keep an assignment log for each class; students record the assignments on it for me daily. Students who are absent can simply refer to the bulletin board for their missed assignments, thereby permitting students to be responsible for their own makeup work.

## Organize File Drawers

For the next part of this system, I use one drawer of a file cabinet for each class. For instance, the top file drawer may be designated for my algebra students, the second file drawer may be for my geometry students, and so on. Each file drawer contains file folders that are labeled a-1 to a-45, b-1 to b-45, c-1 to c-45, and d-1 to d-45. When I first got started I needed 180 file folders for each drawer! I printed labels quickly on the computer and had a few students help me place the labels on the folders. Now, when a-16, the sixteenth assignment in the first quarter, is a worksheet in geometry, it is placed in the geometry drawer in the a-16 folder. When students refer to the bulletin board for a missed assignment, they can find the worksheet for themselves in the appropriate file drawer. The folders can be cleared out at the end of the school year and used again the next year.

## Adapt Ideas to Work for you

These ideas originated from veteran teachers, and with some fine-tuning, they have worked well for me. This organizational system can function for different courses and for classes with relatively large enrollments. If you decide to try any of these organizational ideas or strategies, adapt them according to what works best for you.

## Students' Space

*Janelle Kockler*

At the beginning of each school year, I set up two important areas in my room. The first is a bulletin board labeled "The Wall of Fame," where I place students' outstanding homework assignments or class work. Even at the middle school level, students love to see their work displayed. Students particularly like to show their parents this display on special nights. At the end of each semester, I remove all the papers and sort them into class periods. I then randomly select one paper from each class period to receive the "Wall of Fame Award" at the school awards assembly.

The other area is designated for student materials. This area has shelves that contain such items as scratch paper, facial tissue, graph paper, donated folders, and extra pencils. I tell students that they are welcome to use any items in that area. If a student comes to class and is missing some materials, he or she may go to the materials area and find something they can use. Early in the school year, I send out a list of suggested materials that parents can donate to help fill this area.

## Something I Never Learned in Methods Class: Preventing Invisible Students

Do not let students become invisible in your classroom. Greet them by name when they enter your room. Comment on their accomplishments in sports and clubs. Consider how you might use related information in your mathematics class.

—*Margaret R. Meyer*

## Topic Files as an Organizational Tool for the Classroom

*Susie Tummers*

As it is in almost any career, organization is essential to success in teaching. The organizational tool that has proved to be most beneficial during my career is my system of *topic files.* As a beginning teacher, I collected the work that I did for a particular class and placed all of it in a three-ring binder. Difficulty arose in subsequent years when I attempted to locate a particular lesson or activity. The binder was quite full, and locating specific items became a problem; hence, the topic-file strategy was born.

For each course I teach, I make a list of topics. I place these topics in file folders to start my hanging-file system. Each folder contains activities, lessons, worksheets, and so on, that pertain to that topic. When the time comes to teach a certain topic, I can easily locate the folder that contains a collection of all the information I have on that subject and begin to plan my lesson.

I have found three significant benefits to maintaining topic files. First, the files are an effective organizational tool. All topics are organized, and each file is within quick reach. If an activity or lesson focuses on more than one topic, I can easily make a few photocopies and place the duplicate information in as many topic files as is appropriate. Second, topic files provide a place for me to keep the lessons that colleagues share with me and, in turn, enable me to share ideas with colleagues. I am delighted when teachers come to my classroom asking whether I have a file on a particular topic. Third, the files are a wonderful way to organize the material I receive at the professional development conferences that I attend. When I return from a conference, I simply place photocopies of the lessons and notes in the appropriate files.

Topic files offer an organizational strategy that can work for beginning teachers at all grade levels and in all disciplines. This flexible system can easily be tailored to suit your specific needs, modified to accommodate textbook changes, or moved if you transfer to a new school. This simple idea may seem trivial, but it is one that I wish I had learned earlier in my own career.

A "MUST-DO" LIST

# A "MUST-DO" ORGANIZATION LIST FOR BEGINING MATHEMATICS TEACHERS

*Jane M. Till-Schröder*

Regardless of setting, these general planning guidelines are helpful for all mathematics teachers:

1. Request copies of the student and teacher handbooks immediately after signing your contract. Begin to formulate your own classroom rules using the general school parameters listed in the handbooks.

2. Read and highlight the student handbook for important policies that you and your students will follow on a daily basis.

3. Consider using Power Point to present important classroom information. Print out slides, and use them around the classroom as posters. Alternatively, give a slide show on the first day of classes. Slides may include—
   - classroom rules for behavior,
   - inspirational words,
   - bell schedules, and
   - classroom policies on grading and attendance.

4. Ask what supplies the school provides. These should include pencils, erasers, rulers, lined paper, graph paper, construction paper, transparency films for the classroom and the photocopier, transparency markers, thick permanent markers, and red pens.

5. Create your own storage system, including shelves to organize paperwork for different classes; label the shelves clearly so that students can tell them apart.

6. Have seating charts prepared for the first day of school to send a strong message to any student who is intent on disrupting your class. Make changes throughout the semester as necessary.

7. In the first day of school, hand a blank index card to each student and request any information you think you may need, such as—
   - full name,
   - emergency contact names and telephone numbers,
   - extracurricular activities, and
   - special classroom accommodations necessary to succeed.

8. Post a bell schedule in the room, perhaps near the clock. Having this schedule handy will help you as much as the students.

9. Students are only as organized as their teacher. Take the initiative, and enforce some rules for organization:
   - Require all notes and assignments to be kept in one place.
   - Have students head every paper with their full names, the name of the course, the date, and the assignment.

10. Make sure students record each assignment and corresponding score. This record will help both students and their parents track their progress and, perhaps, stop asking you to do it for them! Use this procedure for each class:
    - Distribute an assignment sheet to be affixed to the inside cover of students' notebooks.
    - Make a folder for each class, and affix the assignment sheet inside the cover to allow students who have been absent to find assignments easily.

11. Keep a large three-ring binder for each subject you teach, complete with the curriculum designed for the class placed in the front. Maintain this binder by following these practices:
    - As you complete each unit, place important masters, keys, handouts, and assessments in the binder.
    - Color-code and flag items that emphasize a particular standard.
    - Keep copies of students' work to demonstrate that standards were met.

# How to Control Make-up Work

*Denise D. Chavis*

I have designed a simple process for make-up work—a four-step procedure that works! Several of my colleagues have adapted it to meet their personal needs. Now my students often make up their work without being prompted. I stay caught up, and I am never stressed at report-card time because of make-up work still outstanding.

## My Four-Step Process

In my classroom, I have a magnetic board that I use by attaching magnets to the back of expandable folders, then affixing the folders to the board. A simple filing shelf from an office supply store or a bulletin board can be used as an alternative. I place four expandable folders on my magnetic board and designate them according to the four labels below.

### 1. "While you were out" forms

In this folder are multiple blank copies of a form I designed to work for my classes. Students fill out the top portion with their name, class period, and date(s) they were absent and return the paper in the second folder. (I like to use colored paper for these forms because the color stands out in the folder when the students return the form later.)

### 2. Mrs. Chavis's mailbox

I check this box during my down time. I fill out the portion directed to the students. My form is divided into three sections—classwork, homework, and tests. I write what was missed in the appropriate sections and return the forms to the students. If the forms are completed during the class period, I give them back to the students directly; otherwise, I place them in the fourth folder. The students then pick up their forms on the next class day.

### 3. Make-up work completed

The students staple their completed make-up work to the form I gave them and place it in my mailbox. I then grade and record their work before placing it in the fourth folder.

### 4. Student mailbox

This box is used to return student work and completed forms. I tell the students to check the box when they expect work coming back to them. I also go through this box periodically and return work that has not been collected.

## Tailoring the Procedure

The difference this simple procedure has made in my paperwork is unbelievable. The most important thing about any system, however, is to make it fit your needs. You will not use it unless it is designed to fit your personal style and the requirements of your students.

# Nobody Can Really Take Your Place, but a Substitute Has to Try: Hints to Make Having a Substitute a Positive Experience

*Nancy Powell*
*Cathy Denbesten*

Not knowing when you might need a substitute can be stressful; however, being organized can lessen the stress and enable you to avoid last-minute preparations. One helpful organizational technique is to create a substitute notebook that can remain in your classroom. Label the notebook clearly, and leave it in a prominent place or make sure that a colleague knows where to find it in the event you are absent. Organize and mark sections to ensure that a substitute can understand and refer to the notebook easily.

## Information for a Substitute Binder

Suggestions for information to include in the substitute binder are listed below.

- A welcome note that thanks the substitute for coming and provides helpful information, such as the handiest place to hang a coat or safely stash a purse, the names and room numbers of teachers in nearby classrooms who are able to answer general or mathematics-related questions, and instructions on where to find needed lesson-plan books

- Classroom rules and regulations, including suggested discipline procedures and locations of detention or discipline forms and hall passes; clear plastic sheet protectors to house copies of these forms in the binder
- A procedure for getting help or contacting the office
- General procedures for fire drills, tornado drills, injuries, and so on
- Forms that need to be filled out every day, and instructions on where to find them
- A chronological list of class periods, including lunch, and times or days that are scheduled for extra duties
- Accurate seating charts with the name that each student commonly uses and attendance sheets that can be carried out of the classroom in an evacuation. Many grade-book software programs include options for printing seating charts with students' pictures, which can be helpful for substitutes.
- Notes for each class that include—
  - ✓ the names of helpful, trustworthy students;
  - ✓ the names of students who have special medical needs or may need emergency care;
  - ✓ the names of non-English-speaking students or students who are enrolled in English as a second language; and
  - ✓ any other important information of which a substitute should be aware
- Clear, understandable lesson plans with instructions about where to find necessary handouts, materials, and supplies for the lesson; notes or keys for worksheets, tests, or quizzes; and an optional activity for each class in the event that students finish their assigned work
- Special activities that can be used any time of the year in an emergency
- A place to leave notes regarding behavior, absences, accomplishments, and questions, by class

## Reminders for Students

Spend a few minutes talking to your classes about your expectations for them when you are absent. Remind students that the reason they are in class is to learn and that they need to use every opportunity to do so whether or not you are physically present. When you return, address any issues documented by the substitute. Do not ignore even minor misbehavior unless you want to hear about more problems the next time you are absent.

## Something I Never Learned in Methods Class: Practice Selective Hearing

Do not comment on everything students say in your classroom. Sometimes it is better to exercise selective hearing. Create a classroom atmosphere that encourages students to treat one another with respect.

*—Margaret R. Meyer*

Notes:

5 SECTION

# EQUITY

As a beginning teacher, you will need to think about how to achieve equity in your mathematics classroom. An important resource to consult in your efforts is the Equity Principle in *Principles and Standards for School Mathematics* (NCTM 2000). This principle offers a broad outline for implementing practices that serve all students. This section complements that discussion by introducing several important ideas to consider for the students who may enter your classroom from diverse backgrounds. In particular, be aware of the issues outlined in the following paragraphs.

## Evaluate Your Own Beliefs and Biases

Evaluating your own beliefs about, and biases toward, school mathematics as a discipline and about your students' learning of school mathematics is essential. To begin your reflection, you might ask yourself the following questions and try to answer candidly:

- How do I learn mathematics?
- What are some ways in which my students learn mathematics most effectively?
- Do I believe that only certain students in my classroom can learn mathematics? If so, why do I hold those beliefs?
- What expectations do I hold for my students?
- Do my beliefs and biases limit my students' opportunities to learn mathematics?
- How might I change my beliefs and biases to ensure that I promote opportunities for my students to learn mathematics?

Most likely, your responses to these types of questions will dictate your daily teaching practices and habits. As you become more aware of equity issues and more committed to making necessary accommodations to "promote access and attainment for all students" (NCTM 2000, p. 12), your own beliefs may change; in turn, you will influence your students' beliefs through the expectations you convey to them about their learning of mathematics.

## Consider the Commitment

Beginning and experienced teachers alike should realize that maintaining educational equity requires a substantial commitment. You may be called on to invest considerable time and energy in activities to ensure equitable conditions in your classroom and school. Whether alone or with colleagues, you should brainstorm strategies to ensure fairness in your teaching and implementation of school policies and curricula. You may have to collaborate with colleagues to develop better understandings of the "strengths and needs of students who come from diverse linguistic and cultural backgrounds, who have specific disabilities, or who possess a special talent and interest in mathematics" (NCTM 2000, p. 14). Certainly, you should explore the literature about equity issues to continue your professional growth in this area and increase your sense of dedication to your students' mathematical achievement.

## Teach Students, Then the Subject

Initially, this perspective may seem awkward, but keep in mind that regardless of the subject or the students' backgrounds, you are first teaching *individuals*. Establishing a good rapport with your students and setting high academic expectations for them are important. Further, you should vary the ways in which you conduct mathematics instruction to ensure that students relate to your teaching and perform to your levels of expectation.

## Learn More about Equity Issues

Examine the readings in this section to learn more about teaching mathematics to females, students who have special needs, and culturally diverse students, including English-language learners. These readings may shed light on your understanding of the strengths and needs of students who look, speak, or behave differently than you did as a student or than you do now as an adult.

# Mathematics Instruction That Works for Girls

*Abbe H. Herzig*
*Rebecca Ambrose*
*Olof Steinthorsdottir*

Mathematics instruction has traditionally catered to a small segment of the population. We encourage you to consider some of the following suggestions gleaned from educational research for making your mathematics instruction accessible to all students, especially girls.

## Vary instructional strategies

Discourse has become an important element in mathematics classes, but it does not always have to be conducted in a whole-class setting. Students who are less assertive or less confident about speaking publicly—as many girls are—are less likely to participate actively in whole-class discussions. Working cooperatively with their peers in small groups gives students opportunities to engage actively in mathematics in a less threatening environment.

Too often, mathematics classes become the site of competition, with students being encouraged to be the fastest and the brightest. Some students, including many females, find this kind of competition distasteful and resist participating in it. Group work can create a supportive environment in which the emphasis is on learning together instead of surpassing other students. Many girls prefer this environment.

Although individual seatwork or whole-class discussions might work best for certain types of lessons, group work is often an effective way to engage all students in performing high-quality mathematics tasks. Different students flourish in different environments, and by varying the types of work you do in class, you give more students opportunities to do their best.

## Focus on understanding and problem solving

When mathematics is taught as an abstract set of rules and recipes for solving various categories of problems, students learn to solve those specific problems but may not be able to use that knowledge in other contexts. Some research suggests that even though all students, boys and girls, prefer to understand the mathematics they are studying, girls will distance themselves from mathematics when they do not understand it, whereas boys are less bothered by a lack of understanding. When students learn mathematics with understanding, they are able to use that knowledge as a foundation for further learning.

Instead of focusing on correct answers, make problem-solving strategies the primary emphasis of your instruction. By examining your students' strategies, you will start to learn what they understand and can plan future instruction accordingly. Encourage your students to explain their reasoning and to actively listen to, question, and learn from, one another. They will discover that mathematics makes sense, and more students will maintain their interest in the subject.

## Make mathematics meaningful

Students who do not see the use for mathematics are less likely to stay interested in it. Girls seem to be particularly sensitive to issues of relevance in mathematics. Problems that are set in meaningful contexts can be excellent motivation for students who need to see the relevance of what they do. The contexts chosen should be interesting and relevant to all students in your class, both girls and boys. Too often, mathematics problems focus on topics that interest boys, without a balance of problems that appeal to girls. As you get to know your students, you can choose problem contexts that are most meaningful to them; doing so can also be a great way to engage students of diverse ethnic backgrounds. You can also ask students to write problems for others in the class to solve. This exercise can be a helpful way to involve students, learn how they think about mathematics, and develop meaningful problem contexts.

## Use a deliberate strategy for calling on students

In whole-class discussions, boys tend to raise their hands faster than girls. If you call on the first person who responds to a question, you leave out those students who require more time to think about the question and formulate responses. Research shows that girls still tend to be less assertive and may need time to develop confidence before raising their hands. Experiment with different strategies for getting more students involved in class discussions. Always give students plenty of time to think before calling on anyone. You might try silently counting to ten before choosing a student to respond to a question. You might also call out names for responses instead of choosing students who have raised their hands, to ensure that everyone has equal opportunities to participate.

### Examine the types of feedback you give students

Some research shows that teachers interact with students in different ways—most likely, without realizing it. In responding to students' work, teachers sometimes ask boys more sophisticated and challenging questions than they ask girls. Because most teachers are probably unaware of this behavior, we should all continually reflect on how we talk to our students. Be sure to reward all students' progress equally and ask them all questions that challenge them to think more deeply about mathematics.

### Maintain high expectations for all students

Research shows that when teachers have different expectations for different students, their interactions with students reflect those expectations. If a teacher does not believe that a student has high potential—because of the student's race, gender, disability, or ethnicity, for example—then the teacher may be satisfied with a lower level of performance from that individual than from a student who the teacher thinks is more capable. Further, students are likely to live up, or down, to the teacher's expectations. In your classroom, maintain high expectations for all students because all are capable of learning mathematics with understanding. •

## Contributing to Young Women's Learning of Mathematics

*Suzanne Gosch*

As a beginning teacher of mathematics, you should be aware of certain practices within your classroom that might present barriers to young women's learning mathematics. You can incorporate numerous suggestions into your instruction to ensure that both females and males have equal opportunities to succeed in mathematics. A few suggestions include—

- being aware of your body language, that is, tone of voice or gestures that may communicate negative messages about mathematics to females;
- being careful not to talk down to girls in the mathematics classroom;
- calling on male and female students equally;
- counteracting the feeling of isolation for girls who enjoy mathematics by introducing them to mathematics teams or mathematics clubs;
- discussing issues of mathematics anxiety and learned helplessness with females who may be struggling; and
- using nontraditional assessments and cooperative learning, both of which have been shown to help females develop their mathematics skills.

Countless programs across the country have also been created to increase females' interest in mathematics; you should encourage your female students to participate in such programs when possible. Moreover, you should coach parents to support their daughters' mathematics education and to encourage them to reach their greatest potential in the field of mathematics. Learning mathematics must be deemed, by society, essential for all students. Given that you have taken these steps, your female students' self-confidence and assurance in mathematics are likely to rise significantly; thus you will have positively contributed to their learning of mathematics.

## Bringing High Expectations to Life in an Urban Classroom

*Ido Jamar*
*Vanessa R. Pitts*

We often hear that one of the most challenging issues facing beginning teachers is classroom management. Although this statement is true for teachers in any setting, it is especially true for beginning teachers in urban classrooms. Clearly learning cannot take place in a classroom that is out of control, but are rules and regulations all that are needed to foster an optimal learning setting for urban students? Haberman (1991) found that "urban schools that serve as models of student learning have teachers who maintain control by establishing trust and involving their students in meaningful activities [therefore] … discipline and control are primarily a *consequence* of their teaching and not a *prerequisite* condition of learning" (p. 293).

## Maintaining High Expectations for All

Unfortunately, whether a teacher engages students in meaningful, challenging tasks often depends on how the teacher perceives the abilities of the students. Too often, minority students have been victims of teachers' low expectations, with the result that students' achievement has mirrored those expectations. High expectations manifested through both words and deeds are necessary if *all* students are to reach high levels of mathematics achievement.

## Conveying Expectations to Students

Many facets of teaching can send students a subtle message that tells them that you hold high expectations for them. For example,

a) by providing opportunities for students to understand concepts prior to learning rules, you make it clear that you know they *can* understand the content and that it is understandable;
b) by using students' existing knowledge as building blocks for new knowledge, you let them know that they already have the necessary foundation needed to learn; and
c) by expecting students to be active participants in their own learning, you make it clear that they are to take responsibility for their own learning.

Although these pedagogical practices are consistent with current reform recommendations, they also create a space in which academic excellence can become the norm, not the exception, for minority students (Ladson-Billings 1994).

# Research Findings Involving English-Language Learners and Implications for Mathematics Teachers

*Sylvia Celedón-Pattichis*

To help you build support for English-language learners in your classroom, the following paragraphs describe some terminology and research findings that address language-acquisition issues affecting mathematics teaching and learning.

## Learning Language through Silence

English-language learners need a *silent period* (Baker 2001), that is, time to acquire the new language without necessarily producing it. This period often lasts two to five months and can be as long as a year. This silence should not be interpreted as unwillingness to participate in mathematics classroom activities. Teachers might use cooperative learning or pairing of students to ensure that English-language learners are exposed to the new language in nonthreatening ways.

## Acquiring Social and Academic Language

English-language learners usually acquire basic interpersonal communication skills, or the *social language*, on the playground, through peer conversations, and in other informal settings during their first two years in a new culture. However, they need five to seven years or more to use the academic language required in educational content areas. The ability to use the language needed to perform in an educational context is referred to as *cognitive academic language proficiency* (Cummins, as cited in Baker [2001]).

Basic interpersonal communication skills are usually learned in *context-embedded* situations. For example, the act of talking face-to-face and using nonverbal gestures with students gives them instant feedback and clues to support spoken language. In contrast, cognitive academic language proficiency tends to be *context reduced*, meaning that no clues are available to support comprehension. Mathematics is often taught abstractly, but for English-language learners, the subject must be put into context with, for example, diagrams or the students' own definitions, especially when students are being introduced to new vocabulary and word problems.

## Developing the Mathematics Register

The *mathematics register* (Halliday 1978), or the specific language used in mathematics, is often problematic for English-language learners. Contrary to popular belief, mathematics is not a universal language. The mathematics register contains many words that have different meanings from what students initially expect. Educators should distinguish between the meanings of words that overlap in the everyday use of English and in mathematics. Teachers can use many strategies to develop the mathematics register in their students, including reinterpreting words in the everyday language, such as *point, reduce, carry, set, power,* and *root.*

## Making Appropriate Placements

The Third International Mathematics and Science Study indicates that the U.S. eighth-grade mathematics curriculum is at a seventh-grade level in comparison with that of other countries (United States Department of Education 1997). Thus, care should be taken in placing English-language learners in mathematics classes to avoid repetition of content. Follow these procedures when determining placement of students from other countries:

- Use test results as supplementary information only.
- Ensure that students have a translator who can help them understand examination instructions and that the translator explains the purpose of the test results.
- If a textbook is available, compare and contrast the curriculum used in the student's previous country with that of the United States.
- Negotiate possible placements by talking to teachers, parents, the student, and counselors. If the student needs to take a mathematics course that is offered at a higher grade level, then the school district should make that option available.
- Ask for previous school contact information from the student or the family to learn what mathematical skills the student demonstrated in his or her previous school.

## Sharing Learners' Cultures in the Classroom

To establish a positive learning environment, mathematics educators should consider the linguistic and cultural experiences that English-language learners bring to the classroom. For example, if a student uses a different algorithm for division, demonstrate it to the class to create a learning moment for everyone. Finally, research students' backgrounds and seek the help of bilingual teachers or specialists in English as a second language to help your students adjust to their new culture.

# Meeting the Challenge of Special Needs Students in the Middle School Classroom

*Julia A. Sliva and Mary Fay-Zenk*

Many topics developed in upper elementary and middle-grades mathematics classes involve the transfer of lower-level concepts to a new context. This transition can be very difficult for many special needs students. Whereas some students may not have mastered the prerequisite skills, others may have difficulty understanding applications of the material or making important connections with other mathematics topics. For example, although some students can solve for an unknown in a proportion, they cannot set up a proportion correctly in a given situation. Consider the following scenario:

*Tommy, an energetic young person who speaks with confidence and participates in class discussions intelligently on most days, still has a lot of trouble taking tests or quizzes and completing his homework. He is easily distracted by other students, is socially adept, and does not seem to have the patience or perseverance needed to finish assignments. Tommy says that he likes mathematics, and from class work you are sure that he understands quite a lot. Why, then, does he do so poorly on written work?*

The following are some issues that may interfere with a special needs student's ability to learn mathematics:

- Understanding such concepts as *first* and *greater than*
- Remembering information
- Mentally shifting from one task to the next
- Developing key perceptual skills, for example, spatial relationships, size relationships, and sequencing
- Developing fluency with mathematics facts
- Maintaining positive attitudes toward learning mathematics
- Selecting appropriate strategies to solve problems
- Developing abstract reasoning.

Uncovering the root of Tommy's difficulties may be much more challenging than realizing that he needs to "sit still." Tommy may have difficulty decoding, with the result that he may not be able to read about an unfamiliar situation with any significant level of comprehension. He may have perceptual deficits such that information as it is presented on the page of the text is overly confusing, or he may be so graphically challenged that copying information accurately is a significant struggle for him. His hyperactivity may be related to the stress of not being able to read, and thus, avoiding the situation by being busy with other activities could be an easy way for him to cope.

As a teacher, your challenge is to try to discover what attitudinal, perceptual, and processing issues cause your student to withdraw in the classroom. What would help promote a positive attitude toward learning mathematics? What would increase a student's participation in learning and his or her willingness to develop strategies for problem solving?

One of the most difficult questions that new teachers encounter each year is, How does one appropriately meet the needs of such special learners as Tommy? Our students bring a continually expanding, wide range of ability levels and experiences to the classroom. Therefore, the task of meeting all learners' needs can seem daunting, even to an experienced teacher. Although as a new teacher, working with experienced mathematics teachers and other school personnel can help you learn very valuable strategies to help you meet the needs of all your students. Accessing a student's Individualized Education Plan (IEP) through her or his special education teacher can be a great place to start! In addition, this colleague can not only propose effective strategies for accommodating a particular student's disability but often give you valuable insight into her or his learning style.

Many strategies that support special education students have been found to also be effective for all students. These approaches may include the following:

- Preteaching vocabulary
- Using concrete manipulatives to teach abstract concepts
- Using multiple representations
- Using real-world applications
- Modeling problem solving
- Promoting a positive attitude toward learning mathematics
- Assisting students in their development of strategies
- Increasing students' exposure to important material

Thus, the strategies that you take the time to develop for your special needs students will prove beneficial for all learners in your classroom.

## Tips for Teaching Culturally Diverse Students

*Joan Cohen Jones*

As a beginning mathematics teacher, I taught in a culturally and ethnically diverse district in the southeastern United States. My students were recent immigrants from Africa, Asia, and Mexico. Our soccer team had representatives from seventeen countries, and our PTA newsletter was printed in nine languages. During my first year, I remember becoming frustrated with a Cambodian student who would not look at me when I spoke to him. Only later did I learn that in his culture, this behavior was a sign of respect rather than disrespect. This experience helped me realize that my lack of knowledge about students' cultural backgrounds hampered my ability to communicate with them effectively.

While I learned more about my students' cultures, I also developed effective strategies for teaching mathematics to diverse students. The following tips are drawn from my experiences as a teacher, student-teaching supervisor, and teacher educator. These strategies can work for all students, but they are specifically designed to create positive experiences for culturally diverse students. They are appropriate for all grade levels and can easily be adapted to other disciplines.

### Check for existing knowledge

When planning for instruction, decide what concepts students need to know to learn the new material. Include checks for existing knowledge at the beginning of your lessons. If necessary, restructure your lessons to develop the requisite knowledge.

### Listen to what students say

Listen to students' comments, questions, and responses. Only by listening carefully can you learn what your students know, what they misunderstand, what is important for them to learn, and what are the best ways for them to learn.

### Question students to reinforce learning and build students' confidence

Ask questions to reinforce students' learning, for example, "How do you know?" "Who knows how to find the answer?" "Will that approach always work?" "Have you seen other problems like this one before?" "Can you find a pattern?" Such questions encourage students to give explanations, search for connections, and rely on themselves and their peers, building self-reliance, cooperation, respect, and confidence.

### Increase wait time during classroom discourse

Increasing your wait time during classroom discourse enhances your listening and questioning skills and improves communication. When listening to students' questions, for example, wait until they are finished speaking instead of cutting in after you think you know what is being asked. Pause for several seconds before trying to answer the question yourself or asking someone else to answer it. Then repeat the question to make sure that you understand what is being asked. When a teacher really listens, students are encouraged to ask the questions they might otherwise keep to themselves. Being given extra time to answer can also be helpful to female students and students whose first language is not English.

### Respect students' abilities and competence

Demonstrate respect for your students' abilities and competence by giving them high-level intellectual tasks that require complex processing and critical thinking. Doing so conveys your confidence that they can master the material.

### Become familiar with and respect students' cultures

Find out as much as possible about each culture represented in your class. Parents and other family members are wonderful resources to help you learn about your students' backgrounds. If possible, spend some time in the neighborhood in which your school is located. Shop at the local grocery store, participate in neighborhood festivals and celebrations, eat at local restaurants, and become familiar with community activities. This practice can enhance your cultural understanding.

### Be reflective

Become a reflective practitioner. That is, monitor, review, and revise your practices, instructional choices, and methodology consistently to ensure that you are being fair and open-minded, providing high-level tasks for your students, and connecting your students' cultural backgrounds with new concepts. An important step in reflection is to think about your own cultural heritage and understand your own biases. Reflection lets you review your own actions in light of what you know about yourself, including your strengths, weaknesses, biases, and prejudices. Some teachers find that keeping a journal, making records of class discussions, or videotaping their classroom instruction is helpful in this process.

### Offer students choices

Offer your students instructional choices. Student choice is especially important for culturally diverse students, who may have learning styles that differ from those of their peers in the dominant culture. When developing classroom activities, offer choices in assignments and types of assessments; give students the opportunity to work independently, in pairs, or in small groups; and allow students to respond in oral or written form, individually, in teams, or as a class.

### Enjoy the challenge

Teaching mathematics to diverse students can be challenging but extremely rewarding—both personally and professionally. I hope the ideas discussed here are helpful as you begin your career in education.

Notes:

6 SECTION

# SCHOOL AND COMMUNITY

Although you will undoubtedly find yourself very busy in your first year of teaching, you will need to find time to reach out to parents and the surrounding school community. In addition to learning more about your students, you will discover that establishing relationships with parents and the community will yield the resources and support required to be successful. When teachers, parents, and the community act together, students benefit in numerous ways. This cooperation often leads to increased rates of attendance, improved academic achievement and graduation rates, and generally better attitudes about school.

Parents are your students' first teachers and your greatest resource. Parents can provide you with valuable information about their children's interests, attitudes, environment, habits, and aptitudes. Parents can be your staunch allies and partners, encouraging their children to persevere and achieve in mathematics. To tap this important resource, often, you only have to reach out and invite parents into a partnership. The same approach also works for members of the community and other significant adults in your students' lives. Encouraging and valuing their contributions strengthen and reinforce their involvement.

As you plan for the school year, develop a strategy to encourage and nurture these vital partnerships. You can start on a small scale by—

- holding regular parent meetings and conferences;
- recruiting classroom volunteers to serve as tutors, guest speakers, and general classroom helpers;
- establishing a school-home connection with homework;
- publishing a parent newsletter; and
- welcoming parents, grandparents, or other involved adults to school.

In this section, you will find other suggestions about how to begin this process, including tips on communicating with parents, sharing your ideas with them, and involving them in the mathematics education of their children. Begin formulating your plan by reviewing these suggestions and reflecting on how they might work for you. Develop goals for parental and community involvement, and jot down notes about how you might accomplish these goals. Talk with your mentor and colleagues about their strategies, and seek out additional resources, such as *Involving Families in School Mathematics: Readings from "Teaching Children Mathematics," "Mathematics Teaching in the Middle School," and "Arithmetic Teacher,"* available from NCTM (Edge 2000). Most of all, develop a positive relationship with the families of your students. Doing so is a rewarding experience and well worth the investment of your time.

# Calling Home: Keeping in Contact with Students' Families

*Laura Brader-Araje*

As a new mathematics teacher, I knew that taking on too much and trying to be perfect at everything would leave me only heartache and stress. For this reason, I decided to be selective in my goal setting for the beginning year. Included in my plan was my hope to develop a system to track communications with the parents and guardians of my students. I knew that the primary method of communication between teachers and students' families at that time was the telephone. Thus, keeping track of incoming and outgoing calls to students' homes would be the basis of any system I would devise.

## The System

I decided on an index card system, which may sound simple or even archaic, but what a gem this scheme was for me! I taught five classes, and in my system, each class was coded with a different colored card. For example, period 1 was yellow, period 2 was blue, and so on. Each student was initially allocated a single note card, listing his or her name, parents' or guardians' names, and home and work telephone numbers. I alphabetized the cards in a large index card file.

## My Proactive Approach

To initiate positive relationships with all these parents, I took a proactive approach; I decided to go to the parents before they came to me. To do so, I tried to make at least five phone calls a day, leaving messages on machines more often than actually hearing another person on the other end of the line: "Hello, my name is Laura Brader, and I am your daughter's new mathematics teacher. I am calling to introduce myself and to touch base. If you ever need anything, please give me a call at school. I look forward to meeting with you!" I could keep my home number unlisted because I was readily available before, during, and after school hours most of the time.

## The Benefits

I maintained my calling routine in between returning calls from parents who had specific concerns throughout the year, because the benefits of the system clearly justified the amount of time involved. The index-card system proved useful over the course of the school year for a variety of reasons, including the following:

- I used the index cards to document the date and time of a call, the person with whom I spoke if someone was available, and the purpose of the call. This record gave me evidence to address problems and to use in team meetings with parents.
- The phone calls helped head off many potential problems because they gave me the chance to mention homework attempts, classroom behavior that was becoming disruptive, concerns about absences, and other issues.
- The system prompted me to call all parents over time, more than once, whether or not I had a specific reason to do so. The calls showed that I cared, and I did, about all my students, not just the infamous ones whoes parents receive phone calls each year from all their child's teachers.
- When problems arose and team meetings were called, the documentation on the index cards allowed me to clarify my approaches toward resolution.
- I created the "luxury," as one team member called it, to transform the dreaded phone call home into an opportunity to tell parents that their children had done well on a project or showed great improvement in a particular area. I found that the parents of students who do not cause problems rarely receive timely, positive feedback about their children; my phone calls provided this praise.
- The initial round of phone calls and messages gave me insight into which students erased messages from teachers before their parents came home and which students did not. I came to understand these students' experiences when teachers had "called home" in the past.

## Addressing Administrators' Concerns

In my experience, new teachers often find that their actions—and inaction—are under constant scrutiny.

Administrators may ask, "Have parent phone calls been returned?" "Are guardians being notified of excessive absences?" "Do students know why calls are being made to their homes?" The index cards, kept in a box by the telephone, allow new teachers to answer this barrage of questions they will receive in their first nine months on the job.

## Closing Thoughts

Calling home keeps me in contact with the parents or guardians of my students and seems to help head off trouble before it starts. Documenting these interactions provides the backup or proof that teachers need in this age of accountability. Finally, a new teacher's attempts to reach out to parents when school starts and to keep in contact with them throughout the year creates a positive impression of the school and bridges the traditional gap between teacher and community.

# Parents in the Mathematics Classroom

*Denise Edelson*

Get your students' parents involved in your mathematics instruction. Do not assume that they are aware of the latest research, methods, or standards in teaching. Most parents have only their personal elementary mathematics experiences as a reference point; chances are these experiences did not include manipulatives, authentic instruction, or an integrated curriculum. Parents should have some point of connection with the program that you are implementing with their children. Accordingly, invite parents into your classroom for a time of mathematics learning, and give them a taste of the components of your thoughtful, standards-based program. The following are some suggestions.

## Suggestions to welcome parents into the classroom

- Parents will come to your classroom if they feel comfortable. Reassure them. I suggest that you invite parents into your classroom during a school day and let them sit with their child. Some parents may feel very inadequate and worry that they will be unable to complete even a primary-grade mathematics assignment on their own. Reassure them that they will not be judged on their work. Feed them. Food is comforting and friendly. I send out party invitations and present an "M & M" mathematics lesson at my first parent gathering. Everyone happily eats candy at the end of the lesson. I make sure to serve good treats at all my mathematics events.
- Parents will do almost anything because they love their child. In thinking of activities that are interesting and entertaining to parents, I get concerned that parents might become bored from, or disdainful of, doing primary-grade work with their child. But I have discovered that parents are just happy to be attending the event with their child, because by doing so, they show their child that they care. Now, although I still present activities that I think the parents might enjoy, I have stopped worrying about entertaining them.
- Parents want you to love their child, too. Parents sometimes hear only unflattering things about their child at school—what he or she needs to do to improve or to behave appropriately. They may wonder whether the teacher even likes their child. An informal mathematics gathering is an opportunity to talk about a child's strengths and abilities. Compliment individual students on their patterning-block pictures, unit-cube patterns, and bean sorting. Their parents will beam and be more likely to calmly accept suggestions for improvement at another meeting, because you have shown that you care and have demonstrated that you can see strengths in their child.
- Parents want to be more involved, but they do not always get good information from their children about what is happening at school. A face-to-face meeting is your chance to show parents that hands-on mathematics is not just "play" despite what their children have told them after school. Your students' parents will be more supportive of your program when they have had opportunities to understand how play translates into mathematical experiences.
- Parents want their children to succeed, in spite of what parents may do or say. They will take afternoons off from work (Friday afternoons worked for me.) to attend an important school activity.
- Parents learn in similar ways that children learn. Do not just tell parents about the wonderful materials that you are using in your classroom. Get out those counters or any other manipulatives, and give the parents a hands-on lesson; read them an entertaining children's mathematics trade book, show a portion of

a good mathematics video, set up interest centers that children can explain to their parents, and post information about pieces-per-serving and calorie count near the cookies (mathematical connections that can be carried into the home). While the children eat cookies, visit with the parents and joyfully talk about...mathematics! •

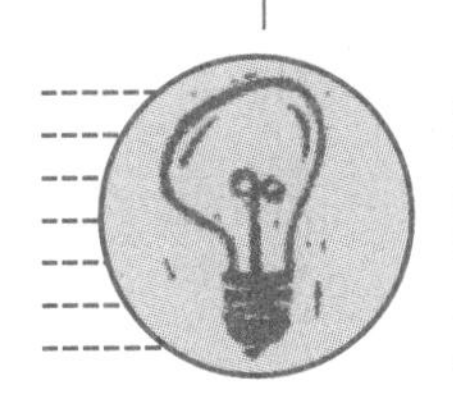

## Sharing Your Principles and Standards with Parents

*Mark W. Ellis*
*Robert Q. Berry III*

We have found that most parents, when well informed about a teacher's motives and methods, are supportive of efforts to promote higher-order thinking, conceptual understanding, mathematical communication, and authentic problem solving in the classroom. Many parents see these efforts as a refreshing change from the focus on algorithms they experienced in learning mathematics, which they may remember as a series of unrelated procedures that mysteriously transformed numbers into answers.

To generate parental support, you need to communicate the mathematical goals that you have set for students and the ways in which you are helping students achieve these goals—in essence, your own principles and standards. To help parents become knowledgeable about these goals, you may wish to send home a parent newsletter or make a presentation at a PTA meeting or school open house. You might also create an opportunity for parents to share in the mathematical experiences that their children are receiving in your classroom by hosting a family mathematics night.

These outreach efforts are important because parents who have experienced mathematics through traditional procedure-oriented instruction often have questions about teaching strategies that you may apply in your classroom, for example, cooperative work, student inquiry, manipulative modeling, and writing assignments. Also, consider sharing with parents research findings that support NCTM Standards–based instruction. Parents will see firsthand how Standards-based practices enhance students' understanding and enable them to attach meaning to mathematical concepts and ideas.

## Things I Never Learned in Methods Class: Parents and Support Staff

Do not call parents with only bad news. Call them to report children's improvements or hard work, also. Tell them how much you enjoy having their child in your class.

Do not take the school secretaries or custodians for granted. Remember their names, and greet them with smiles. Take the time to express your appreciation for the work they do.

—*Margaret R. Meyer*

Notes:

# ONLINE RESOURCES FOR THE BEGINNING TEACHER

The Internet is one of the best resources available to you as a beginning teacher. As you probably already know, numerous Web sites are available for practically any instructional topic in which you are interested. A few sites serve as central clearinghouses, providing links to related sources of information. The Web sites listed below have been active for a while; they are fairly stable in terms of accessibility and are kept up-to-date with the latest educational developments. As part of your professional growth plan, explore these sites to search for information on particular topics or issues pertaining to mathematics teaching and learning. Make an attempt to visit at least one every month, if for no other reason than to learn what is happening in the educational world outside your classroom.

- For professional growth opportunities, visit the National Council of Teachers of Mathematics (NCTM) site at http://www.nctm.org. NCTM membership is required to gain access to some of the features.
- For instructional ideas, visit the Math Forum Web site at http://mathforum.org/. In particular, you might want to explore "Teacher2Teacher" for answers to your questions about teaching mathematics. You can browse the archives, search the frequently asked questions, or submit a question of your own.
- For professional publications related to the implementation of curriculum and instruction, visit the Association for Supervision and Curriculum Development site at http://www.ascd.org.
- For assessment information, view the searchable online journal *Practical Assessment, Research and Evaluation* at http://www.ericae.net/pare.
- For current research publications, browse the education category of the National Academies Press Web site at http://www.nap.edu.
- For a variety of publications related to education, visit the Eisenhower National Clearinghouse site at http://www.enc.org.
- For research and resources on general topics pertaining to various aspects of education, visit the U.S. Department of Education on the Web at http://www.ed.gov.

**Notes:**

# REFERENCES


Baker, Colin. *Foundations of Bilingual Education and Bilingualism.* Philadelphia, Pa.: Multilingual Matters, 2001.

Crannell, Annalisa. "Collaborative Oral Take-Home Exams." In *Assessment Practices in Undergraduate Mathematics,* edited by B. Gold, S. Z. Keith, and W. A. Marion, pp. 143–45. Washington, D.C.: Mathematical Association of America, 1999.

Edge, Douglas, ed. *Involving Families in School Mathematics: Readings from "Teaching Children Mathematics," "Mathematics Teaching in the Middle School," and "Arithmetic Teacher."* Reston, Va.: National Council of Teachers of Mathematics, 2000.

Feiman-Nemser, Sharon. "What New Teachers Need to Learn." *Educational Leadership* 60 (May 2003): 25–29.

Flores, Alfinio. "Electronic Technology and NCTM Standards." 1998. http://mathforum.org/technology/papers/papers/flores.html (accessed 12 April 2001).

Haberman, Martin. "The Pedagogy of Poverty versus Good Teaching." *Phi Delta Kappan* 73, no. 4 (December 1991): 290–94.

Halliday, Michael Alexander Kirkwood. *Language as Social Semiotic.* Baltimore, Md.: Edward Arnold, 1978.

Kroll, Diana Lambdin, Joanna O. Masingila, and Sue Tinsley Mau. "Grading Cooperative Problem Solving." In *Emphasis on Assessment,* edited by Diana V. Lambdin, Paul E. Kehle, and Ronald V. Preston, pp. 50–57. Reston, Va.: National Council of Teachers of Mathematics, 1996.

Ladson-Billings, Gloria. *The Dreamkeepers: Successful Teachers of African American Children.* San Francisco: Jossey-Bass Publishers, 1994.

Little, Judith W. "The Mentor Phenomenon and the Social Organization of Teaching." *Review of Research in Education* 16 (1990): 297–351.

National Council of Teachers of Mathematics (NCTM). *Principles and Standards for School Mathematics.* Reston, Va.: NCTM, 2000.

Renard, Lisa. "Setting New Teachers Up for Failure…or Success." *Educational Leadership* 60 (May 2003): 62–64.

United States Department of Education. *Introduction to TIMSS: The Third International Mathematics and Science Study.* Washington, D.C.: United States Department of Education, 1997.


Notes:

Two additional titles appear in the
**Empowering the Beginning Teacher of Mathematics**
series
(Michaele F. Chappell, series editor):

***Empowering the Beginning Teacher of Mathematics in High School***
Edited by Michaele F. Chappell, Jeffrey Choppin, and Jenny Salls

***Empowering the Beginning Teacher of Mathematics in Elementary School***
Edited by Michaele F. Chappell, Jane F. Schielack, and Sharon Zagorski